Tomart's Price Guide to
Lowe and Whitman
PaperDolls

by Mary Young

Color Photography by **Tom Schwartz**

Masthead designed by **Fred Blumenthal**

Black and white photography by the author.

Other books by the author:

Paper Dolls and Their Artists — Book I
Paper Dolls and Their Artists — Book II
A Collector's Guide to Paper Dolls - Saalfield, Lowe, Merrill
A Collector's Guide to Paper Dolls - Second Series
A Collector's Guide to Magazine Paper Dolls
Paper Dolls and Their Aartists — Revised Edition
Tomart's Price Guide to Saalfield and Merrill Paper Dolls

TOMART PUBLICATIONS
Division of Tomart Corporation
Dayton, Ohio

DEDICATION

To Our Newest Grandchild
Grace Carol

ACKNOWLEDGEMENTS

My deepest thanks to all of you who made your books available for picturing. The credit given next to your paper dolls seems very little in comparison to the value of your contributions. Without you, this book could not have been written. I appreciate each and every one of you! Those who helped in other ways are **Esther Bermann, Elizabeth Boardman, Marjorie Buxton, Kay Dirschell, Mary Domen, Natalie Eckenfels, Pam Glenn, Bonnie Gurzenda, Pam Kalla, Pam Kleinedler, Sharon Rogers, Estelle Spyes, Kathy Winn**, and **Wynn Yusas**.

Extra special thanks go to **Peggy Ell, Edith Lowe, Virginia Crossley, Rosalie Eppert**, and **Audrey Sepponen** who provided much needed information at a moment's notice.

As always, my husband **George** endured the time I spent immersed in writing with his usual good humor and patience, always ready to help in proofreading and other ways.

Much appreciation goes to **Bob Welbaum, Nathan Zwilling**, and **T.N. Tumbusch** on the staff at Tomart, and to **Tom Tumbusch** for making it all possible.

Prices listed are based on the author's experience. They are presented as a guide for information purposes only. No one is obligated in any way to buy, sell, or trade according to these prices. Condition, rarity, demand and the reader's desire to own determine the actual price paid. No offer to buy or sell at the prices listed is intended or made. Buying and selling is conducted at the reader's risk. Neither the author nor publisher assumes any liability for any losses suffered for use of, or any typographical errors contained in, this book. All value estimates are presented in U.S. dollars.
All paper dolls are from the author's collection unless otherwise noted.

Library of Congress Catalog Card Number: 92-60232

ISBN: 0-914293-58-3 (0-914293-19-2)   Manufactured in the United States of America

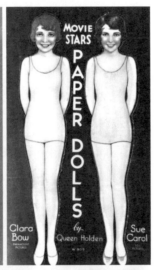

INTRODUCTION

This book covers the paper dolls published by the Lowe and Whitman Publishing Companies. All known original paper dolls and their reprints are listed and all of the listed originals are pictured. Reprints having the same dolls as the original are not pictured but are listed.

Because many paper doll collectors also collect books and box sets of paper toys and stand-ups of dolls, animals, toys, etc., these are also included in the general lists when they are known. It should be understood, though, that they do not contain paper dolls with outfits, and they will not be shown in the picture sections.

The books in the picture sections and general lists are listed in numerical order according to the publishers' number which appears on the book cover. If the book has no number, then it appears at the end of the list. Most numbers have 3 or 4 digits. Occasionally you may find a number on a book reading as follows: 2723:15. The 15 is not part of the number but merely the publisher's code for the price of the book; i.e., 15¢.

The general lists will serve as a check list for all the original and reprint books. Concerning the reprints, the list also includes, in most cases, the number of the original book from which it came.

Price Guide

The prices in this book are based on mint, uncut, original paper dolls. Reprints that are almost identical are just slightly lower in price. In the case of a reprint of a celebrity book where the dolls have been redrawn, the price is drastically reduced. There are a few rare cases of a reprint having a higher value than its original. Such is the case of a few non-celebrity books by the Lowe Company that were reprinted with redrawn dolls and made into celebrity books.

Cut sets are usually half the price of the uncut set providing that all the dolls and outfits are included and the pieces are in very good condition. If, however, any dolls, or outfits are missing, bent, mended, or torn, etc., the price decreases accordingly.

The prices in this guide were largely derived from many detailed studies of different sales lists of paper dolls and notes taken at paper doll conventions where paper dolls were sold. Information and knowledge from other collectors and personal judgement were other influencing factors.

Paper Doll Collecting

Readers unfamiliar with the recent hobby of paper doll collecting may wonder at the interest of avid paper doll collectors. Although paper doll collecting may not seem an inspiring pastime to some, collectors have found the historical and cultural value of the paper doll to be not only fascinating, but enjoyable as well.

Adults who remember the happy hours of enjoyment paper dolls brought them as a child need only to meet or hear of another paper doll collector to get caught up in this hobby. Even those who never played with paper dolls become infatuated with the hobby, some just because they are interested in the dress designs of past eras.

Paper doll collecting has been growing rapidly in recent years, getting a good start in the 1960's. Collectors range in age from small children to those in their 80's and 90's. Not everyone collects the same type of paper dolls. As with other hobbies, everyone has different interests. Some collect only the celebrity paper dolls, while others like only the non-celebrity. Some like only the uncut paper dolls, while others like them cut out. There are some collectors who collect only the paper dolls that appeared in magazines or newspapers and then there are others who, like myself, collect just about all the different kinds of paper dolls there are!

Paper Doll History

No one knows for sure just how paper dolls were originated. One popular belief is that they began as pantins (jumping jacks) in Europe. Pantins were dolls made of cardboard with arms and legs that moved when a string attached to the parts was pulled.

Paper dolls during the 19th century were of famous dancers and of the opera star, Jenny Lind, plus many of the non-celebrity kind, all of which are very rare and hard to find now. In the late 1800's the Raphael Tuck Publishers in England produced many beautiful series of paper dolls. These were distributed in the United States through their New York office.

Advertising paper dolls were used by many companies in the late 1800's to further the sale of their products. The ONT Thread Company, Lion Coffee, McLoughlins Coffee, Ceresota Flour, and Pillsbury Flour are a few examples. Many times the child could send away for a complete series of paper dolls after receiving the first doll with the product. This type of advertising continued into the 1900's and even today has not completely disappeared.

Around the turn of the century popular women's magazines started to include a page of paper dolls. The most popular were *Ladies' Home Journal*, *Good Housekeeping*, *Pictorial Review*, *McCall's*, *Delineator*, and *Woman's Home Companion*.

From the beginning, commercial paper dolls could be bought in books, boxes, envelopes, or folders; but the type of paper dolls most familiar to us today is the paper doll book which features the dolls on the cardboard covers and the clothes on the inside pages. This type of book made its appearance in the late 1920's and the early 1930's and sold for a very reasonable price of either five or ten cents. This meant that children everywhere in the country could enjoy paper dolls with their saved pennies. Even during the depression years these books sold well and parents could buy paper dolls for birthdays and holidays for a fraction of what "real" dolls sold for.

The various companies that produced the paper dolls were very encouraged by the excellent response to the paper doll books and started publishing more and more of these books. The war years of the early 1940's saw the biggest surge of paper dolls which was never to be equaled again. Every company seemed to try to outdo one another with beautiful books of non-celebrity and celebrity paper dolls. Some celebrity books from those years are Alice Fay, Claudette Colbert, Judy Garland, Greer Garson, and Rita Hayworth, not to mention the child stars of Margaret O'Brien and Gloria Jean. Shirley Temple was a teenager by now; however, many paper dolls of Shirley as a child appeared in the 1930's and one set as a teenager in the 1940's. There were paper dolls of the stars of radio programs such as "Hour of Charm," "Glenn Miller," and "Benny Goodman" plus paper dolls from entire movies such as *Ziegfeld Girl* and *Gone With The Wind*. One book called **Hollywood Personalities** featured the stars from the Bing Crosby movie *Holiday Inn*.

Aside from paper doll books, children of the 1930's and 1940's also enjoyed paper dolls that appeared in the comic sections of their local newspapers. *Blondie*, *Brenda Starr*, and *Jane Arden* were very popular at that time, and children eagerly awaited the Sunday comics to get their new paper dolls.

Paper dolls were still produced in good quantities in the 1950's. However, the advent of television was being felt and children were not quite as interested in sitting down to read a good book or cut out paper dolls as before. From the 1960's on, there has been a steady decline in the amount of paper dolls produced. However, you can still find some in your local stores in lesser quantity. The more recent paper dolls in the celebrity category are those from the television shows "The Brady Bunch," "The Partridge Family," and "The Waltons." Paper dolls of "real" dolls were made as far back as the 1930's. The Dy-Dee-Doll is one example. Recently there have been dozens of this type of paper doll, with the Barbie Doll well out in front.

It is often asked how one goes about starting a paper doll collection. If you are interested in starting a general collection, begin by buying the paper dolls available in the stores now. Someday these too will be sought after. If you want to concentrate on paper dolls from a special era or one particular type of paper doll, make a point to visit flea markets, antique shows and garage sales. You may also want to check friends and neighbors. They might have a box of paper dolls from their childhood stored away in the attic just waiting for you!

1046 Clothes Crazy

On The Subject Of Reprints

Very often an original paper doll book was reprinted after it was published. Sometimes the book was reprinted exactly like the original, but more often it was changed somewhat. The following typifies the many different ways an original may have been reprinted:

The reprint and original may be exact duplicates.

The reprint may be an exact duplicate but with a different price.

The reprint may be the same but with no price on the book at all.

The reprint may have fewer pages.

The reprint may be printed on a lesser grade of paper.

The reprint may have a new background on the covers; dolls and inside pages remain the same.

The reprint may have completely new covers, dolls redrawn while inside pages remain the same.

The reprint may have fewer dolls than the original.

The reprint may have a different title than the original.

The reprint may be a box set while the original was a book.

The reprint may be a book while the original was a box set.

The reprint may be made up of dolls from two, three or even four different paper doll books.

The reprint may have dolls and outfits of reduced or enlarged size.

1044 College Girls

The reprint may not have die-cut dolls though the original book did.

The reprint may have coloring pages added or subtracted.

The reprint may have a reprint of its own.

The reprint may be an exact duplicate with a different trade name.

One common type of reprint is better known in the business as a jobber book. The Lowe Company used the trade name of "Abbott" for their jobber books. The dolls and clothes in these books were the same as the original except that usually two or more pages were dropped and the covers were made of very lightweight cardboard, rarely die-cut. The jobber book was usually placed on the market at the same time as the original main line book; however, these less expensive books did not go to the big chain stores but rather to small toy and variety stores, drug stores, train stations, and bus depots.

The following is an example of reprints with redrawn dolls. Number **1046 Clothes Crazy** was published in 1945 by the Lowe Company. In 1946, a new book was produced with new dolls redrawn in the same poses as the original dolls. The original clothes inside the book were retained. This book was **#1044 College Girls**. Then in 1961 a third book, **#2760 Dolls**, was designed with a third set of dolls in the outlines of the original dolls but retained the same clothes.

Because of the great number of reprints with redrawn paper dolls by the Lowe Company, the author has grouped as many

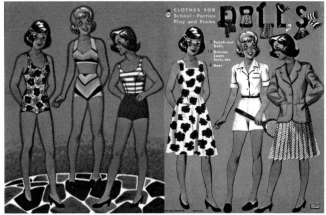

2760 Dolls

as possible together on two of the color pages; they are described at the end of the index.

A reprint may appeal to the collector just as much as the original book, especially if it contains added material or if the dolls are redrawn by a favorite artist. Many collectors try to collect both the original books and all their reprints.

The above information on reprints refers only to reprints published by the original company within a few days, months, or years of the original. In recent years a number of paper doll books have been reprinted by companies other than the initial company and can still be bought today. These reprint books give the collector an opportunity to own copies of some beautiful early paper dolls which are now hard to find uncut. The books will usually be noted on the cover or on the wrapping that they are a reprint. It would be good if these new reprints also had some printing on the backs of the dolls to allow a collector to distinguish the new from the old in case the dolls are cut out, but unfortunately only a few do have this feature. However, collectors will find that most of the dolls, when cut out, will have a distinguishable white color on their backs.

Anyone with paper doll questions or information is encouraged to write the author at P.O. Box 9244, Wright Brothers Branch, Dayton, Ohio 45409. Please enclose a SASE.

THE SAMUEL LOWE PUBLISHING COMPANY

Mr. Samuel Lowe was the founder and first president of the Samuel Lowe Publishing Company. He was born in 1884, and began his career in New York City working at the Henry Street Settlement House which compared with the Hull House in Chicago. Later, when Mr. Lowe moved to Racine, Wisconsin, he worked at the Central Association, a social service organization. His love for children and interest in children's books prompted him in 1917 to take a job in Racine with the Western Printing and Lithographing Company (now Western Publishing Company) to develop a line of children's books which eventually became incorporated into the Whitman line.

Mr. Lowe was the originator of the Big Little Books. One day at Western he found some scrap chunks of paper left over from cutting and trimming. Mr. Lowe walked around with these chunks for two weeks saying what nice books they would make for children's small hands. But he couldn't think of a use for them until he realized that a comic strip square was exactly the same size as these chunks of paper. So he went down to the Chicago Tribune and arranged to have Chester Gould (the creator of Dick Tracy) do the first Big Little Book which naturally was called **The Adventures of Dick Tracy, Detective** #707, copyright 1932. Later, Mr. Lowe also met with Walt Disney and another Big Little Book was born - **Mickey Mouse** #717, copyright 1933.

Mr. Samuel Lowe 1884-1952
Founder of the Samuel Lowe Publishing Company
Painting by George Pollard

Twenty-three years after Samuel Lowe came to Western, he decided that it was "now or never" to try his hand at his own publishing company. So he left Western and started his own firm a few miles away in Kenosha, Wisconsin. Among the company's early successes were their novelty books and novelty handle books consisting of several storybooks tied together with a ribbon or cord which then ran through a cardboard handle. A few years later, the company launched its very popular line of Bonnie Books. These are discussed in some detail near the end of the picture section for Lowe.

The Lowes were friends of the publishing firm of Raphael Tuck & Sons of England. In the early 1940s Mr. Lowe was named an honorary director to the Tuck Board of Directors. During wartime, the Lowe Company published some books for the Tucks and sold some of the Tuck greeting cards and calendars here in the United States.

Mr. Lowe was president until his death in 1952. He was succeeded by his wife Edith Kovar Lowe. Mrs. Lowe had also worked for Western, starting in 1924 and leaving with her husband in 1940, the year the Lowe Company was founded. There are five Lowe sons; Samuel Jr., James, Jonathan, Peter, and Richard. Three of the sons, Jonathan, Peter, and Richard, have held positions in the company.

Mrs. Lowe also authored many children's books in the 1930's for Whitman under her maiden name of Edith May Kovar and also under her pen name of Mary Windsor. She continued to write many children's books for the Lowe Company in the years that followed.

Paper dolls were a big part of the Lowe Company's main line of children's books from the beginning. In the early years, they were drawn by such noted artists as Rachel Taft Dixon, Pelagie Doane, Doris and Marion Henderson, and Fern Bisel Peat, and more recently by Queen Holden, George and Nan Pollard, Jeanne Voelz, Elsie Darien, and Harriet Hentschel Struzennegger.

Since about the mid-Fifties, many of the Lowe books were copyrighted by the James and Jonathan Company, an affiliated firm which designed the books. Other trade names used at various times by the company were John Martin's House, Abbott Publishing, Angelus Publishing, Lolly Pop Books, and Faircrest.

The Lowe Company ceased publishing in 1979.

521 Little Cousins 1940 $30 - 50   **521 Twinkle Twins** 1944 $25 - 30   **522 Twinkle Twins** 1944 $25 - 30   **523 Bab and Her Doll Furniture** 1943 $25 - 30 Right: inside front cover

    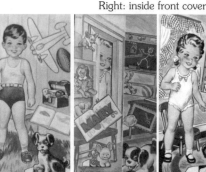

523 Farmer Fred 1943 $25 - 30 Right: inside front cover   **523 Janie and Her Doll** 1943 $25 - 30 Right: inside front cover   **523 Tom and His Toys** 1943 $25 - 30 Right: inside front cover   **523 Mary and Her Toys** 1943 $25 - 30 Right: inside front cover

Cut set from **521A Playmates** 1940 $15 cut set, $40 uncut book

955 Dr. Kildare Play Book $20 - 25   **955** inside page

The Samuel Lowe picture section includes paper dolls from the James and Jonathan Company and paper dolls using the trade names of Abbott Publishing and John Martins' House.

958 Career Girls 1950 $35 - 50

968 Square Dance 1950 $25 - 30 Date in Lowe records

990 TV Tap Stars 1952 $25 - 35

1022 Tina and Tony 1940 $45 - 55

1021 The Baby Show - 25 Dolls 1940 $60 - 100 Right: Inside front and back covers

1023 Sally and Dick, Bob and Jean 1940 $50 - 65 Right: Inside front and back covers

1024 Polly Patchwork and Her Friends 1941 $50 - 75

1027 In Our Backyard 1941 $75 - 100

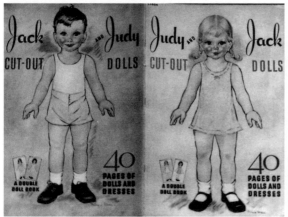

1024 Judy and Jack, Peg and Bill 1940 $60 - 100 A Double Doll Book. Right: Inside front and back covers

1025 The 8 Ages of Judy 1941 $150 - 250

1025 Turnabout Dolls 1943 $50 - 65 Right: Reverse of inside front and back covers

8

1026 Beauty Contest 1941 $75 - 100

1026 Dude Ranch 1943 $35 - 50 Right: Inside back and front covers

1028 Playhouse Paper Dolls 1941 $50 - 65 contains reversible clothes. Right: Inside front and back covers

1028 Girls in the War 1943 $85 - 150 Turnabout Doll Book. Right: Inside front and back covers

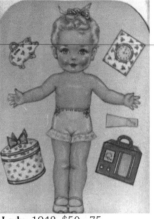

1029 Clothes Make A Lady 1942 $50 - 75
Some books dated 1941. Right: Inside Front Cover

1030 Annie Laurie 1941 $50 - 75

1030 Little Women 1941 $50 - 75

1030 The Five Little Peppers 1941 $50 - 75

1040 King of Swing and Queen of Song 1942 $300 - 400 Benny Goodman & Peggy Lee. Right: Inside back and front covers

1041 Glenn Miller - Marion Hutton 1942 $300 - 400 Turnabout Doll Book. Right: Inside back and front covers

1042 Junior Prom 1942 $40 - 60

1042 Pat The "Stand-Up"
Doll 1946 $35 - 50
Date in Lowe records

1044 Blue Feather Indian 1944 $50 - 75
Silver Cloud Indian on back cover

1043 Betty Bo-Peep 1942 $50 - 85 Billy Boy Blue on back cover

1043 The Bride Doll 1946 $75 - 100

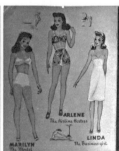

1044 Me and Mimi 1942 $50 - 75
Includes small paper doll book of Mimi

1045 Wee Wee Baby
Doll Book 1945 $35 - 50

1045 Career Girls 1942 $50 - 75

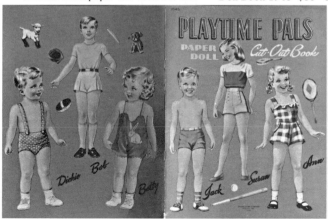

1045 Playtime Pals 1946 $25 - 30

1046 Clothes Crazy 1945 $35 - 50 Date in Lowe records

1048 Girls in Uniform 1942 $75 - 125

1048 The Turnabouts Doll Book 1943 $35 - 50

1049 Lollypop Crowd 1945 $40 - 60

1049 Hollywood Personalities 1941 $300 - 400

1056 Down on the Farm $30 - 40

1057 Playhouse Paper Dolls 1947 $25 - 35 Patterned after the earlier #1028 Playhouse; new dolls and clothes were drawn.

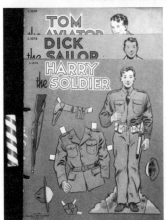

Tom, Dick, and Harry were sold together as a set.

1074 Tom the Aviator 1941 $60 - 85

12

1074 Dick the Sailor 1941 $60 - 85

1074 Harry the Soldier 1941 $60 - 85

1077 Little Bear to Dress 1942 $35 - 50
1077 Little Dog to Dress 1942 $35 - 50
1077 Little Pig to Dress 1942 $35 - 50
1077 Little Kitten to Dress 1942 $35 - 50
These four books are tied together and sold as
one. Date is from the Lowe Catalog. See color
pages for individual photos.

1242 Cinderella Steps Out 1948 $60 - 75 Right: Inside front cover

1246 Fashion Previews 1949 $45 - 60

1251 New Toni Hair-Do Dress-Up Dolls 1951 $75 - 90

1248 Let's Play House $25 - 35 The furniture in the book originated
with #7502 Doll House, a box set published in 1943. There is no date on
this book, but it is shown in the 1949 catalog.

1252 Rockabye Babies 1952 $25 - 35

1253 Prom Home Permanent 1952 $40 - 60

1254 The Bobbsey Twins 1952 $60 - 75
Date in Lowe records

1256 Rosemary Clooney 1953 $100 - 175 **#2585 Gloria's Make-Up** was redrawn in 1953 and became Rosemary Clooney. Six of eight pages of clothes from Gloria were retained in the Rosemary Clooney books.

1283 Cuddles and Rags 1950 $50 - 65 Date in Lowe records

1284 Toni Hair-Do Cut-Out Dolls 1950 $60 - 75
The first edition has 6 pages, the second only 4 pages.

1286 Cowboys and Cowgirls 1950 $30 - 50
Date in Lowe records

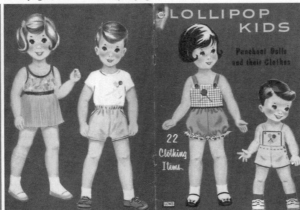

1829 Playmates 1961 $12 - 15
Date in Lowe records
1831 Lollipop Kids 1961 $15 - 18
Date in Lowe records
1832 Lots of Dolls 1961 $12 - 15
This book uses clothes from **#1057 Playhouse**, but the dolls are completely new.

#1829 is redrawn from **#1056 Fun on the Farm** and #1831 is from **#1049 Lollypop Crowd**.

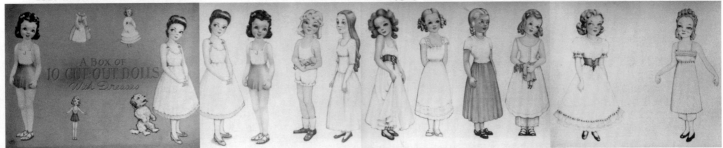

1842 A Box of 10 Cut-Out Dolls With Dresses $150 - 250 This beautiful box set was drawn by Helen Page. Each doll has her own little folder complete with clothes and accessories. The folders are of a fine quality paper which accentuates the delicate pastels used by the artist. In addition to the dolls in the folders, the same dolls were also printed on sheets of cardboard. This set was published using one of the Lowe Company trademarks **John Martin's House** and is in the 1945 and 1947 catalog. See color page 82 for these dolls with their folders.

1885 Little Dressmaker Doll Book - Patty 1966 $10 - 15
1886 Little Dressmaker Doll Book - Lucy 1966 $10 - 15
1887 Little Dressmaker Doll Book - Janie 1966 $10 - 15

2110 Santa's Band 1962 $10 - 12 Right: Inside center pages

2403 Baby Doll 1957 $25 - 40 Right: Doll

2404 Three Little Maids From School Are We 1957 $35 - 45

2405 Janet Leigh 1957 $75 - 100

2406 Patti Page 1957 $75 - 100

2407 The Bob Cummings Fashion Models 1957 $75 - 90

2411 Patience and Prudence 1957 $50 - 65

2422 Keepsake Folio - Mimi Doll 1964 $35 - 40
2423 Keepsake Folio - Emily Doll 1964 $35 - 40
2424 Keepsake Folio - Trudy Doll 1964 $35 - 40

2483 Junior Misses 1958 $40 - 50 A folder-type book that opens from the middle.

2488 Patti Page 1958 $75 - 100

2493 Bride and Groom Dolls 1959 $45 - 65

2523 Kay and Kim with Clothes to Trim 1956 $30 - 40

2560 The Honeymooners 1956 $200 - 350

2561 Goldilocks and the Three Bears 1955 $30 - 40

2562 Here Comes the Bride 1955 $40 - 60

**2563 His and Hers - The Guest
Towel Dolls** 1955 $25 - 30

2569 Rosemary Clooney 1956 $100 - 200

2574 Paint Betsy's Sunday Best
1955 $15 - 20 Date in Lowe records

2585 Gloria's Make-Up 1952
$50 - 75

2610 Sally's Dress A Doll
Storybook 1952 18 - 25

2713 Rosemary Clooney 1957 $100 - 175 This is a reprint of #2585 Gloria's Make-Up with redrawn dolls. Gloria was redrawn in 1953 for #1256 & became Rosemary Clooney! The dolls were redrawn again for this book. Six of eight pages of clothes from Gloria were retained in these Rosemary Clooney books.

2713 Mopsy and Popsy 1971 $12 - 15

2714 Dollies Go 'Round the World 1971 $12 - 15

2717 When We Grow Up 1971 $12 - 15

2718 Here Comes the Bride 1971 $20 - 25

2720 Dress Annabelle 1972 $12 - 15

2721 Doll Friends 1972 $12 - 15

17

2722 Little Dolls 1972 $12 - 15

2723 A House Full of Dolls and Clothes 1972 $12 - 15

2724 Now and Then Paper Dolls 1973 $12 - 15

2725 The Happy Family 1973 $12 - 15

Courtesy Audrey Sepponen

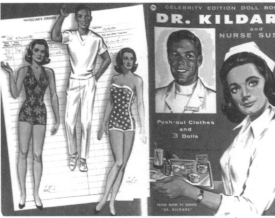

2740 Dr. Kildare and Nurse Susan 1962 $45 - 60

Susan dolls came from **#4913 Airline Stewardess** with some clothes. Dolls were redrawn. The Kildare doll came from **#2407 Bob Cummings** with some clothes. The doll

2743 Two Teens 1963 $12 - 15

was redrawn. There are two different front covers: blue background and signed by Pollard, and an unsigned red background.

2749 Baby Anne Smiles and Cries 1964 $20 - 25

Courtesy Audrey Sepponen

2750 Alive-Like Dolly Elizabeth 1963 $25 - 35

2751 Big Doll Betty 1960 $15 - 20

2751 Alive-Like Dolly Bonnie 1963 $25 - 35

2760 Dotty Doll Book 1968 $12 - 15
A new doll drawn in the figure of **#3905 Mimi**.

2762 Vicky 1967 $20 - 30
Doll is new, clothes are from **#2751 Bonnie**.

2763 Dolly Gets Lots of New Clothes 1967 $20 - 30
Doll is new, clothes are from **#2750 Elizabeth**.

2764 Archie's Girls 1964 $40 - 55
Dolls are redrawn from **#2483 Junior Misses**.

2764 Pixie Doll and Pup 1968 $25 - 35

2766 Front and Back Dolls and Dresses 1964 $12 - 15

2784 Little Girls 1969 $12 - 15

2785 Brenda Lee 1961 $50 - 60 Dolls have been redrawn from **#2407 Bob Cummings**, some clothes are new.

2785 Sally, Sue and Sherry 1969 $12 - 15

2904 Sew-On Sherry Ann's Clothes
1960 $18 - 25

2906 Betsy Dress A Doll Storybook 1960 $18 - 25
2915 Wendy Dress A Doll Storybook 1960 $18 - 25
Each page is a different outfit.

3727 Jeannie and Gene 1975 $12 - 15

3730 Polly Pal 1976 $12 - 15

3947 Cindy - Storybook
Doll 1964 $25 - 30

3921 Patti Doll Book 1961 $18 - 25
With clothes from the **#3903** dolls.

3903 Alive-Like Sally Ann 1960 $10 - 15
3903 Mary Ann 1960 $10 - 15
3903 Betty Ann 1960 $10 - 15

3905 Mimi From Paree 1960 $12 - 15

4171 Gabby Hayes 1954 $60 - 80
A Bonnie Jack-In-The-Box Book

The Samuel Lowe Publishing Company Bonnie Books

Bonnie Books were some of the most unique storybooks ever made. They were small (8 x 6-1/2 inches) and included Pop-up books, Television books with a TV screen that changes pictures as a dial is turned, Push Along books of trains, fire engines, etc., with wheels that turn, books of paper dolls and other cut-out books with scissors attached to the cover. Also part of the Bonnie Book line were books with boxes of crayons or paint sets attached. Another novelty was a baby's book with a rattle attached to the front cover. Then there were Bonnie spinner game books, progressive games with the spinner showing through the pages as the game progressed, and Bonnie Merry-Go-Round books which formed individual stage-like scenes portraying the story's characters and scenery when opened. There were even Bonnie Books with jigsaw puzzles inside, and books of other games and puzzles. The list is endless. Each new catalog illustrated new and unusual Bonnie Books.

4207 Penny's Party 1952 $35 - 60
A Paper Doll Story Book - Bonnie Book

4219 Bye Baby Bunting 1953 $35 - 50
Paper Doll and Story Book - Bonnie Book

4263 Sherlock Bones 1955 $15 - 20 A Jiggle-head Bonnie Book with costumes on each page. The doll's head shows through the front cover of the Jiggle-head books.

4264 Dolly Goes Round The World 1955 $15 - 20

 **4266** Cover

 **4265** Cover

 **4267** Cover

4268 Cover

4266 Betty Plays Lady 1953 $20 - 25 **4265 Billy Boy** 1953 (also 1955) $20 - 25
Jack-in-the-Book Bonnie Book   Jack-in-the-Book Bonnie Book

4267 Popsy 1953 $20 - 25 Jack-in-the-Book Bonnie Book

4268 Trinket 1953 $20 - 25 Jack-in-the-Book Bonnie Book

4282 Little Sugar Bear 1952 $20 - 25 Jack-in-the-Book Bonnie Book

4281 Circus Time 1952 $20 - 25 Jack-in-the-Book Bonnie Book

4283 Dolly Takes a Trip 1952 $20 - 25 Jack-in-the-Book Bonnie Book

4284 Cookie the Rabbit 1952 $20 - 25 Jack-in-the-Book Bonnie Book

4284 and **4343** Inside pages

4343 Captain Big Bill 1956 $20 - 25 Jack-in-the-Book Bonnie Book

4298 Jungletown Jamboree $15 - 20 Spin-dial Bonnie Book

4299 Masquerade Party 1955 $20 - 25 Spin-dial Bonnie Book

4913 Airline Stewardess 1957 $40 - 60

4344 Alfie 1956 $20 - 25 Jack-in-the-Book Bonnie Book

4345 Timmy 1956 $20 - 25 Jack-in-the-Book Bonnie Book

5901 Suzy 1961 $20 - 25 Alive-Like Face Doll Book

5908 Molly Dolly 1962 $20 - 25 Doll and clothes are new, except for four outfits that are from **#3905 Mimi**.

8902 Lace and Dress Puppy 1975 $20 - 25

8903 Lace and Dress Kitty 1975 $20 - 25

9041 Peggy and Peter 1962 $60 - 75
Big Big Doll Book Right: Back cover

9045 Two Little Girls 1964
$45 - 55 Big Big Doll Book
Dolls are new; clothes are from
#2904 and #2915.

9118 Betsy 1964 Box Set $45 - 55
Doll has animated mouth.

6918 Betsy $45 - 55 Dolls and clothes are same as #9118,
except the doll does not have an animated mouth.

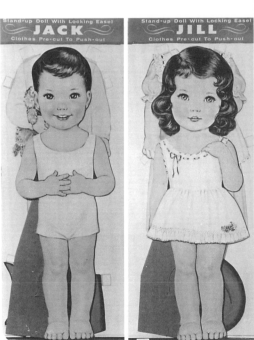  

9301 Jack 1963 $35 - 40
9302 Jill 1963 $35 - 40

9985 Cecelia My Kissin' Cousin
1960 $40 - 50
30 inches tall

9986 Kathy 1962 $40 - 60
30-inch stand-up doll

23

PAPER DOLLS PUBLISHED BY SAMUEL LOWE CO.

This list contains all original and reprint paper doll books and box sets published by the Samuel Lowe Publishing Company plus those published under other Lowe Company trade names as Abbott, James and Jonathan, Bonnie Books and John Martin's House. Also included are a few books that were made for Sears and the Bestmaid Company.

Books of stand-up toys and figures are listed when known.

Reprints will have the number of the original book it is derived from in parenthesis following the title. Brackets without any numbers indicate a reprint where the original is not definitely known. It should be remembered that the reprint is not always the same as the original. For example, sometimes only one or two dolls of the original book were used. Other examples of how reprints may differ from the original are mentioned on page 4.

8	**Playroom Paper Dolls** box of 5 books all dated 1944: #523-1 **Four Jolly Friends**, #523-2 **Susie and Betty**, #525-1 **Three Little Sisters**, #525-2 **Stand-Out Dolls**, #526-1 **Baby Cut-Outs**. Sometimes a 6th book would be substituted: #526-2 **Four Playmates**	
L-24	**Playroom Paper Dolls** box of 5 books all numbered L523: **Tom, Mary, Fred, Bab and Janie**	
L-24	**Playroom Paper Dolls** box of 5 books all dated 1944: #523-1 **Four Jolly Friends**, #523-2 **Susie and Betty**, #525-1 **Three Little Sisters**, #525-2 **Stand-Out Dolls**, #526-1 **Baby Cut-Outs**. Sometimes a 6th book would be substituted: #526-2 **Four Playmates**	
57	**Model Warplanes, Tanks** box, stand-ups (1063, 1069, 1093)	
58	**Victory Girls** box (1048 Girls in Uniform)	
78	**Playroom Paper Dolls** box of 5 assorted paper doll books from the 1940's, books varied as to titles available in stock	
123	**Playmates** (1025 Turnabout Dolls)	
124	**Nancy, Judy, Susan, Betty, Four Cut-Out Dolls** 1942 (1025 The 8 Ages of Judy)	
125	**Playground** 1944 (1025 Turnabout Dolls)	
125	**Four Playmate Cut-Out Dolls** (1021 Baby Show)	
126	**Big Round-Up** (1026 Dude Ranch)	
L126	**Dude Ranch** 1944 (1026)	
127	**Sally and Her Twin Brother Dick** 1944 (1023)	
127	**The Playtime Twins** (1023 Sally, Dick, Bob and Jean)	
128	**The Twins Bob and Jean** 1944 (1023)	
128	**Ann and Betty** (521A and 521 Little Cousins)	
129	**Clothes Make a Lady** 1943 (1029)	
130	**Betty, Jane and Dick** 1943 (1023 Sally, Dick, etc & 1022 Tina and Tony)	
131	**Tina and Tony** (1022)	
132	**Cut-Out Kids** (523 Janie & 521 Twinkle Twin boy)	
132	**My Big Dolls** (1025 Turnabout Dolls)	
132	**Playground** (1025 Turnabout Dolls)	
133	**Cut-Out Fun** ( )	
133	**My Big Dolls** 1944 (1025 Turnabout Dolls)	
140	**Over 80 Turn-About, Stand-Up Soldiers** 1943	
141	**Over 80 Turn-About, Stand-Up Sailors** 1943	
142	**Judy and Mary** (1025 Turnabout Dolls)	
142	**Junior Prom** 1942 (1042)	
143	**Peggy and Carol** (1025 Turnabout Dolls)	
143	**Betty Bo-Peep** (1043)	
L144	**Me and Mimi** 1942 (1044)	
144	**Blue Feather - Silver Cloud** (1044)	
145	**Ten Little Playmates** 1944 (521A)	
145	**Girls in Uniform** 1943 (1048)	
146	**Cut-Out Dolls for Fun and Play** (521 Little Cousins)	
148	**Carol and Jerry** 1944 (1022 Tina and Tony)	
148	**The Outdoor Girls** (1048 The Turnabouts)	
149	**Sue and Tom** 1944 (1022 Tina and Tony)	
228	**Peter and Prue** box (1021 Baby Show)	
254	**Mammoth Box of Dolls and Dresses** box of assorted paper doll books	
294	**Nine Doll Box** (521A)	
313	**Rosemary Clooney** coloring book w/paper dolls on covers, no outfits	
521	**Little Cousins** 1940	
521	**Twinkle Twins 4 Years Old** 1944 (date in Lowe records)	
521	**Sunbonnet Sue** 1943 (1029)	
521A	**Playmates** 1940	

522	**Twinkle Twins 10 Years Old** 1944 (date in Lowe records)	
522	**Sonny and Sue** 1940 (1022 Tina and Tony)	
522	**Debs, A Pressed Board Doll** 1943 (1045 Career Girls)	
522	**Debs** (title on back cover) 1943 (1045 Career Girls)	
522	**Grown-Ups** 1943 (1045 Career Girls)	
522	**Models** 1943 (1045 Career Girls)	
523	**Playroom Dolls** box, contained the following 5 books; #523 **Tom and His Dog** 1943, #523 **Bab and Her Doll Furniture** 1943, #523 **Janie and Her Doll** 1943, #523 **Mary and Her Toys** 1943, #523 **Farmer Fred** 1943	
523	**Peg and Bill** 1941 (1024)	
523	**Judy and Jack** 1941 (1024)	
523-1	**Four Jolly Friends** 1944 (521A)	
523-2	**Susie and Betty** 1944 (521 Little Cousins)	
523-3	**Three Little Sisters and Their Clothes** (521 Little Cousins & 521A)	
524	**Betsy and Bill** 1943 (1021 Baby Show)	
L524	**Mary is a Real Wood Doll** (521 Little Cousins & 521A)	
L524	**Jane is a Real Wood Doll** (521 Little Cousins & 521A)	
525	**Patty and Pete** 1941 (1022 Tina and Tony)	
525	**Stand-Out Dolls** 1944 (521 Little Cousins)	
525-1	**Three Little Sisters** 1944 (521 Little Cousins & 521A)	
525-2	**Stand-Out Paper Dolls** 1944 (521 Little Cousins & 521A)	
526	**Cut-Out Book with Real Wood Doll** (521A)	
526	**Eight Little Playmates** 1944 (1021 Baby Show)	
526	**My Favorite with a pulp Board Doll** 1942 (521 & 521A)	
526-1	**Baby Cut-Out Dolls** 1944 (1021 Baby Show)	
526-2	**Four Playmates** 1944 (1021 Baby Show)	
527	**Sally and Her Twin Brother Dick** 1943 (1023)	
527	**Jean and Her Brother Bob** 1942 (1023)	
528	**Peter and Prue** 1942 (1021 Baby Show)	
528	**Betsy and Bill** 1942 (1021 Baby Show)	
528	**Betsy and Bill** 1943 (1021 Baby Show)	
529	**War Girls** 1943 (1048 Girls in Uniform)	
535	**On Guard** 1942, stand-ups (1074)	
536	**Build Model Tanks** stand-ups (1065)	
536	**Model Planes** 1943, stand-ups (1069)	
537	**Indian Cut-Out Book** 1943, stand-ups	
537	**Cowboy Cut-Out Book** 1943, stand-ups	
602	**Our Sailors** stand-ups, box (1093)	
602	**Our Soldiers** stand-ups, box (1063)	
613	**Doll House Dollies** 1966 (1832 Lots of Dolls)	
621	**World Famous Kissin Cousin** - life size stand-up doll, 1967 (9985)	
622	**Kathy** life size doll, 1967 (9986)	
724	**A Box of Cut-Out Doll Books** 1944, four books: #521A **Playmates**, #523-1 **Four Jolly Friends**, #525-1 **Three Little Sisters** & #526-2 **Four Playmates**; other books may have been substituted.	
730	**Mopsy and Popsy** 1972, box (2713)	
731	**Jane and Jill** 1972, box (2784 Little Girls)	
731	**Kitty and Puppy Lace-On** 1975, box (8902 & 8903)	
732	**Dressmakers Doll Set** 1972, box (1829 Playmates & 1831 Lollipop Kids)	
733	**Dotty Dolly Lace-On Doll** 1972, box (2760)	
848	**Victory Punch-Outs** stand-ups, envelope	
937	**Baby Sitter** (1025 Turnabout & 1046 Clothes Crazy)	
945	**Dress Alike** (1283)	

945	**Baby Sitter** (1025 Turnabout & 1046 Clothes Crazy)	
946	**Down on the Farm** (1056)	
946	**Dance Team** (990)	
947	**Sweet Sixteen-Vivian, Ruth and Peggy** (1253 Prom)	
948	**Saturday Night Barn Dance** (968)	
948	**Sunny and Sue** (1049 Lollipop & 1252 Rockabye)	
949	**Little Neighbor** (1254 Bobbsey)	
950	**Brother and Sister** (521 & 522 Twinkle Twins)	
950	**Campus Queens** (1046 Clothes Crazy)	
951	**Jamboree Dolls** (968)	
951	**Country Cousins** (1056)	
952	**Wedding Party** (1043)	
952	**Prom Date** (1046 Clothes Crazy)	
953	**Swing Your Partners** (968)	
953	**Hollywood Doll Book** (958)	
954	**Big Sisters in Paris** (958 Career Girls)	
954	**Cowgirls** (1026 Dude Ranch)	
955	**Dr. Kildare Play Book** has paper doll	
955	**TV Style Show** (1246)	
958	**Career Girls** 1950 (date in Lowe records)	
958	**Party Dolls** (990)	
959	**Dolly and Me** (1284)	
960	**Toni Hair-Do Cut-Out Dolls** 1950 (1284)	
961	**TV Star Time** (990)	
962	**Campus Queens** (1253 Prom)	
963	**Twinkle Tots** (1252 Rockabye & 1049 Lollypop)	
964	**Cuddles and Rags** (1283)	
965	**My Favorite Doll Book** (1254 The Bobbsey Twins)	
968	**Square Dance** 1950 (date in Lowe records)	
969	**Hollywood Dolls** (1246)	
970	**Judy and Her Friends** five dolls with clothes to be colored (1046 Clothes Crazy)	
971	**Cowgirls** (1026 Dude Ranch)	
972	**Fairground** (1049 Lollypop)	
973	**Boardwalk** (522 & 521 Twinkle Twins)	
975	**Boardwalk** (522 & 521 Twinkle Twins)	
976	**Fun on the Farm** (1056)	
977	**Boardwalk** (522 & 521 Twinkle Twins)	
977	**Fashion Parade** (1046)	
978	**Polka Party** (968)	
979	**Television Stars** (1246)	
980	**Big Sister** (958 Career Girls)	
981	**Campus Queens** (1046 Clothes Crazy)	
982	**Cowgirls** (1026 Dude Ranch)	
983	**Hollywood Dolls** (958 Career Girls)	
984	**Jamboree Doll Book** (968)	
985	**Wedding Party** (1043)	
986	**Models Doll Book** (1246)	
988	**Look-Alike Cut-Out Dolls** (1284)	
989	**Nora Drake** 1952 (958 Career Girls)	
990	**TV Tap Stars** 1952 (date in Lowe records)	
1009	**Fashion Models** (2407)	
1010	**Twins - Baby Dolls, Jane and Jill** (4219)	
1011	**Teen-Age Dolls** (2483)	
1012	**Sister Dolls - Sherry and Nancy** (2411)	
1013	**Honeymoon Dolls** (2493)	
1014	**Mary Jane** (1283)	
1015	**Baby Sitter** (1025 Turnabout & 1046 Clothes Crazy)	
1016	**Young Couple** (2407)	
1021	**Mary Ann Grows Up** 1943 (1025 The 8 Ages of Judy)	
1021	**The Baby Show - 25 Dolls**, 1940	
1022	**Tina and Tony** 1940	
1022	**Clothes Make A Lady** 1942 (1044 Mimi)	
1023	**Sally and Dick, Bob and Jean** 1940	
1023	**Pressed Board Dolls** 1942 (1048 Girls in

1024 **Judy and Jack, Peg and Bill** 1940
1024 **Polly Patchwork and Her Friends** 1941
1024 **Bumpity Bess** 1943 (1022 Tina and Tony)
1025 **The 8 Ages of Judy** 1941
1025 **Turnabout Dolls** 1943
1026 **Paper Doll's Beauty Contest** 1941
1026 **Dude Ranch Turnabout Paper Dolls** 1943
1027 **The Turnabout Twins** 1943 (1023 Sally, Dick, Bob and Jean)
1027 **In Our Backyard** 1941
1028 **Playhouse Paper Dolls** 1941
1028 **Girls in the War Turnabout Doll Book** 1943
1029 **Clothes Make A Lady** 1942 (some books dated 1941)

The following three books were tied together & sold as one:
1030 **Little Women** 1941
1030 **The Five Little Peppers** 1941
1030 **Annie Laurie** 1941
1031 **Doll Packet** unnamed dolls (4913)
1031 **Doll Packet** unnamed doll (2585)
1031 **Doll Packet** Penny & Sue (4207)
1032 **Doll Packet** three dolls (2404)
1032 **Doll Packet** (2407)
1032 **Doll Packet** Peggy (1253 Prom)
1033 **Doll Packet** (2405)
1034 **Doll Packet** Prudence and Patience (2411)
1034 **Doll Packet** (2406)
1034 **Doll Packet** (2488)
1035 **Doll Packet** (2407)
1035 **Doll Packet** Ruth and Vivian (1253 Prom)
1036 **Doll Packet** (1283)
1036 **Doll Packet** Prudence and Patience (2411)
1040 **King of Swing and Queen of Song** Benny Goodman - Peggy Lee, 1942
1040 **Baby Bunting** two versions, paper dolls from #1021 Baby Show or #1045 Wee Wee Baby
1040 **Girl and Boy** (2407)
1041 **Glenn Miller - Marion Hutton** 1942
1041 **Cutie Doll** (1283)
1041 **Cut-Me Out Paper Dolls** Bab, Fred, Tom, Mary and Janie (523)
1042 **Junior Prom** 1942
1042 **Sally the Standing Doll** (1042 Pat)
1042 **Pat the Stand-Up Doll** 1946 (date in Lowe records)
1042 **Hollywood Glamour** (2407)
1042 **Judy and Mary** 1944 (1025 Turnabout)
1043 **Betty Bo Peep - Billy Boy Blue** 1942
1043 **The Bride Doll** 1946
1043 **Peggy and Carol - Two Big Dolls** 1944 (1025 Turnabout)
1043 **Three Little Maids from School** (2404)
1044 **Me and Mimi** 1942
1044 **Bride Doll Book** (2493)
1044 **Blue Feather - Silver Cloud Indian** 1944 (date in Lowe records)
1044 **College Girls** 1946 (date in Lowe records) (1046 Clothes Crazy)
1045 **Career Girls** 1942
1045 **Wee Wee Baby Doll Book** 1945
1045 **Playtime Pals** 1946
1045 **Going Places with Mary and Bill** (2563)
1046 **Fuzzy Heads** 1942 (1029 for the girl & 1043 Betty Bo Peep for the boy)
1046 **Clothes Crazy** 1945 (date in Lowe records)
1047 **Nancy and Judy** 1942, Susan & Betty on back cover (1025 The 8 Ages of Judy)
1047 **Romance** (1842 A Box of 10 Cut-Out Dolls)
1047 **Maid of Board** (1024 Judy and Jack etc.)
1048 **The Turnabouts** 1943
1048 **Girls in Uniform** 1942
1048 **Lots of Dolls** (1025 Turnabout)
1048 **Circus Punchouts** 1961
1049 **Hollywood Personalities** 1941
1049 **Lollypop Crowd** 1945 (date in Lowe records)
1051 **Merry Go Round** in 1962 catalog, stand-ups
1052 **Fire Department** in 1962 catalog, stand-ups
1053 **Our Town** in 1962 catalog, stand-ups
1053 **Here Comes the Bride** (1043)
1054 **Dude Ranch** (1026)

1056 **Down on the Farm** 1940's
1057 **Playhouse Paper Dolls** 1947 (date in Lowe records)
1061 **Hayride** (1056)
1061 **Jalopy Paper Dolls** (1045 Playtime)
1062 **Playground** (1025 Turnabout)
1062 **Baby Dolls** (1025 The 8 Ages of Judy)
1063 **Bab and Her Doll Furniture** (523)
1063 **Janie and Her Doll** (523)
1063 **Mary and Her Toys** (523)
1063 **United States Soldiers** 1942, stand-ups
1064-1 **Princess** (1242)
1064-2 **Outdoor Girls** (1246)
1064-3 **Sally and Sue** (1042 Pat)
1065 **Model Tanks Construction Kit** circa 1941, stand-ups
1069 **Model Airplanes Construction Kit** 1941, stand-ups
1072 **Janie and Her Doll** (523)
1073 **Bab and Her Doll Furniture** (523)
1074 **Mary and Her Toys** (523)

The following three books were tied together and sold as one:
1074 **Dick the Sailor** 1941
1074 **Tom the Aviator** 1941
1074 **Harry the Soldier** 1941
1076 **Hayride** (1056)
1077 **Jalopy Cut-Outs** (1045 Playtime)

The following four books were tied together & sold as one; no date but pictured in 1942 catalog & listed as new.
1077 **Little Bear to Dress**
1077 **Little Dog to Dress**
1077 **Little Pig to Dress**
1077 **Little Kitten to Dress**
1078 **Baby Buggy** (1025 The 8 Ages of Judy with dolls & clothes redrawn)
1079 **Playground** (1025 Turnabout)
1081 **A Box of Three Doll Books** (1021 Baby Show)
1083 **Here Comes The Bride** (1043)
1083 **Novelty Box of 10 Doll Books** nine from #1021 Baby Show & one from #521 Little Cousins
1085 **Wild West** (1026 Dude Ranch)
1085 **Dude Ranch** (1026)
1085 **Big Round Up** (1026 Dude Ranch)
1087 **The Bilt-Up Book of Mother Goose** 1943, stand-ups
1088 **The Bilt-Up Book of Little Red Riding Hood** 1943, stand-ups
1089 **U.S. Commandos** 1943, stand-ups
1093 **United States Sailors Punch-Out Book** 1942, stand-ups
1226 **Twelve Cut Out Dolls** 1944 (date in Lowe records) box (1021 Baby Show)
1235 **Soldiers and Sailors** box, stand-ups (1093 & 1063)
1241 **Clothes Make a Lady** (1044 Mimi with doll & clothes redrawn)
1242 **Cinderella Steps Out** 1948
1243 **Fashion Cut-Outs with Sturdibilt Dolls** (1046 Clothes Crazy)
1244 **Lots of Sturdibilt Dolls** 1057
1245 **Statuette Dolls** (1048 Turnabouts)
1246 **Fashion Previews** 1949
1248 **Twelve Dolls** box (521A)
1248 **Let's Play House** appeared in 1949 catalog
1249 **Hang Up Your Doll Clothes** (958 Career Girls)
1250 **Down on the Farm** (1056)
1250 **Here Comes the Bride** (1043)
1250 **Model Airplanes** 1943, stand-ups
1251 **Model Tanks** 1943, stand-ups
1251 **New Toni Hair-Do Dress-Up Dolls** 1951
1251 **Schoolmates** (1049 Lollypop)
1252 **Rockabye Babies** 1952 (date in Lowe records)
1252 **Square Dance** (968)
1252 **Cowboy Stand-Ups** 1943
1253 **Indian Stand-Ups** 1943
1253 **Prom Home Permanent** 1952
1253 **Coeds** (1253 Prom)
1254 **We're the Jones Family** (2562)
1254 **The Bobbsey Twins** 1952 (date in Lowe

records)
1254 **Career Girls** (958)
1254 **Farmyard** 1943, stand-ups
1255 **Fairy Princess** (1242)
1256 **Beauty Queens** used dolls of Alice & Trixie from #2560 plus dolls from #2523
1256 **Rosemary Clooney** 1953 (date in Lowe records) originated with #2585 Gloria's Make-Up, six of eight pages of clothes were retained
1256 **Furniture for Living Room, Bedroom and Kitchen!** box, stand-ups
1257 **Baby Parade** (1252 Rockabye)
1257 **The Bobbsey Twins** (1254)
1258 **Gloria's Make Up** (2585)
1258 **Sally the Standing Doll** (1042 Pat)
1259 **Fritzi Ritz** (1251 Toni)
1260 **Annie the Sweetheart Doll** 1956 (1283)
1262 **Baby Doll** 1957 (2403)
1263 **Little Maids Dress Shop** 1957 (2404)
1264 **Two Dolls** box (1045 Career Dolls)
1264 **Pressed Board Dolls** 1942 (1025 The 8 Ages of Judy)
1265 **Service Kit, Land, Sea and Air** 1943, stand-ups (1074)
1266 **Model War Planes** 1943, stand-ups
1267 **Model Tank Construction Kit** 1943, stand-ups
1280 **Diaper Doll** (1045 Wee Wee Baby)
1281 **Fun on the Farm** (1056)
1283 **Cuddles and Rags** 1950 (date in Lowe records)
1284 **Toni Hair-Do Cut-Out Dolls** 1950
1286 **Cowboys and Cowgirls** 1950 (date in Lowe records)
1294 **Three Dolls Made of Wood** box (521A)
1294 **9 Dolls with 32 Pages of Dresses** (521A)
1295 **Twenty Dolls** box (1021 Baby Show)
1319 **Engine 69** coloring book w/stand-ups on back cover
1320 **Fashion Parade** (1046 Clothes Crazy)
1320 **Twinkle Twins** (521 & 522)
1321 **The Twelve Doll Cut-Out Book** (521A & 1021 Baby Show)
1322 **Here Comes the Bride** (1043)
1322 **Playground Paper Dolls** (1025 Turnabout)
1323 **Penny Grows Up** (1025 The 8 Ages of Judy)
1323 **Playmates** (1025 Turnabout)
1323 **College Girls** (1046 Clothes Crazy)
1324 **Cut-Me-Out Paper Dolls** (523 Tom, Bab, Mary, Janie & Fred)
1324 **Twinkle Twins** (521 & 522)
1324 **The Fashion Models** (1048 Turnabouts)
1325 **Playtime Pals** (1045)
1325 **Dolls at Play** (1025 Turnabout)
1325 **Playground** (1025 Turnabout)
1325 **Hay Ride** (1056)
1326 **Holidays Paper Dolls** (521 & 522 Twinkle Twins)
1326 **Jalopy Paper Dolls** (1045 Playtime)
1326 **Big Baby Doll Book** (1045 Wee Wee Baby)
1327 **Baby Buggy** (1025 The 8 Ages of Judy)
1327 **Lollypop Crowd** (1045 Playtime Pals)
1327 **Ann and Betty** (521A & 521 Little Cousins)
1328 **Playground** (1025 Turnabout)
1328 **Brother and Sister** (521 & 522 Twinkle Twins)
1329 **Holidays** ( )
1329 **Princess** (1242)
1329-2 **Outdoor Girls** (1246)
1329-3 **Sally and Sue** (1042 Pat)
1331 **Costume Paper Dolls** (1842 A Box of 10 Cut-Out Dolls)
1331 **Wedding Party** (1043)
1332 **Jamboree Doll Book** (968)
1333 **Campus Queens** (1046 Clothes Crazy)
1334 **Cowgirls Doll Book** (1026 Dude Ranch)
1335 **Hollywood Doll Book** (958)
1335 **Fritzi Ritz** coloring book, paper dolls but no outfits, dolls like #1259
1336 **Models Doll Book** (1246)
1337 **Janet Leigh** coloring book w/paper dolls from #2405
1350 **Fashion Parade Doll Book** 1947 (1046 Clothes Crazy)
1351 **The First Seven Years of Penny** (1025 The 8

1352 Ages of Judy)
1352 **Playtime Pals** (1045)
1353 **The Lollypop Crowd** (1049)
1354 ~~The Outdoor Girls (1048 Turnabouts)~~
1355 **Bride Doll** (1043)
1356 **Blue Feather and Silver Cloud** (1044)
1357 **Judy and Mary** (1025 Turnabout)
1358 **Cut-Me-Out Paper Dolls** (523 Tom, Bab, Janie, Fred and Mary)
1359 **Busy Days** (521 & 522 Twinkle Twins)
1360 **Farmyard** (1056)
1361 **The First Seven Years of Penny** (1025 The 8 Ages of Judy)
1361 **Lots of Dolls** (1057)
1362 **Dude Ranch** (1026)
1363 **Career Girls** (   )
1366 **Look Alike Dolls** (1284)
1367 **TV Star Time** 1955 (990)
1368 **Twinkle Tots** (1252 Rockabye & 1049 Lollypop)
1369 **Cuddles** (1283)
1370 **My Favorite Doll Book** (1254 Bobbsey)
1371 **Campus Queens** (1253 Prom)
1372 **Big 'N' Little Sister** (1284)
1372 **Schoolmates** (1252 Rockabye & 1049 Lollypop)
1373 **Coeds** (1253 Prom)
1373 **Dress Alike** (1283)
1374 **Dotty Dimple** (1283)
1374 **Dance Team** (990)
1375 **Sweet Sixteen** (1253 Prom)
1375 **Party Dolls** (990)
1376 **Big Sister** (1284)
1376 **Sonny and Sue** (1049 Lollypop & 1252 Rockabye)
1377 **Play Days Paper Dolls** (1254 Bobbsey)
1377 **Little Neighbor Paper Dolls** (1254 Bobbsey)
1378 **Teen Queens** (1253 Prom)
1378 **Vacation Time** (1246 & 1026 Dude Ranch for Dolls, new clothes)
1379 **Me and Mimi** (1283)
1379 **Toni Hair-Do Cut-Out Dolls** 1950 (1284)
1380 **Schoolmates** (1049 Lollypop & 1252 Rockabye)
1380 **TV Star Time** (990)
1381 **Dress A Like** (1284)
1381 **Campus Queens** (1253 Prom)
1382 **Play Pals** (1254 Bobbsey)
1382 **Twinkle Tots** (1252 Rockabye & 1049 Lollypop)
1383 **Party Fashions** (990)
1383 **Cuddles** (1283)
1384 **My Favorite Doll Book** (1254 Bobbsey)
1384 **Look A Like** (1284)
1385 **Busy Days** (521 & 522 Twinkle Twins)
1385 **TV Star Time** (990)
1386 **Brother and Sister** (521 & 522 Twinkle Twins)
1386 **Country Cousins** (1056)
1386 **Campus Queens** (1253 Prom)
1386 **Lots of Dolls** (1057)
1387 **Sunny and Sue** (1049 Lollypop & 1252 Rockabye)
1387 **Prom Date** (1046 Clothes Crazy)
1387 **The Outdoor Girls** (1048 Turnabouts)
1388 **Jalopy** (1045 Playtime)
1388 **Dress Alike** (1283)
1388 **Swing Your Partner** (968)
1389 **Big Sister in Paris** (958 Career Girls)
1389 **Here Comes the Bride** (1043)
1389 **My Favorite Doll Book** (1254 Bobbsey Twins)
1390 **Fashion Previews** (1246)
1390 **TV Style Show** (1246)
1391 **Boardwalk Paper Dolls** (521 & 522 Twinkle Twins)
1391 **Brother & Sister** (521 & 522 Twinkle Twins)
1391 **Down on the Farm** (1056)
1392 **Fun on the Farm** (1056)
1392 **Judy and Her Friends** (1046 Clothes Crazy)
1393 **Playhouse Paper Dolls** (1057)
1393 **Fashion Parade** (1046 Clothes Crazy)
1394 **Polka Party** (968)
1394 **Beach Party** (521 & 522 Twinkle Twins)

1395 **Television Stars** (1246)
1395 **Carnival Paper Dolls** (1049 Lollypop)
1396 **Cowgirls, A Real Western Doll Book** 1950 (1026 Dude Ranch)
1396 **Big Sister** (958 Career Girls)
1396 **Big Sisters In Paris** (958 Career Girls)
1424 **Ann and Betty** (521A & 521 Little Cousins) made for Best Maid Co.
1425-4 **Playroom Paper Dolls** box, five books with contents similar to #8 & #L-24; made for Best Maid Co.
1428 **The Twelve Doll Cut-Out Book** (521A & 1021 Baby Show) Best Maid Co.
1442 **Four Playmate Cut-Out Dolls** (1021 Baby Show) Best Maid Co.
1443 **Betty Bo Peep** (1045 Wee Wee Baby) Best Maid Co.
1503 **Get Well Aids** 1956 paper doll coloring book
1513 **Mother and Daughter** (2562)
1514 **Ruth, Joan and Winnie** (2404)
1515 **Mr. and Mrs.** (2560)
1516 **Jet Airline Stewardess** (4913)
1517 **Fashion Models** (2488)
1518 **Wendy** (2915)
1589 **Celebrity Autograph Coloring Book with Paper Dolls of Janet Leigh** 1958 (2405)
1801 **Baby Doll** (2403)
1801A **Little Dressmaker Doll Book** - Betty Jane, 1964 (1831 Lollypop Kids)
1801B **Little Dressmaker Doll Book** - Mary Ann, 1964 (1829 Playmates)
1801C **Little Dressmaker Doll Book** - Sally Lou, 1964 (1831 Lollypop Kids)
1802 **Little School Maids** (2404)
1802 **Playroom Paper Dolls** box, contents similar to #8 & #L-24, five books
1804 **Patti Page** 1958 (2406)
1805 **Janet Leigh** 1958 (2405)
1806 **Rosemary Clooney** 1959 (2585)
1807 **Patience and Prudence** 1959 (2411)
1808 **The Three Cheers** (1253 Prom)
1809 **Here Comes the Bride** (1043)
1810 **Bob Cummings Fashion Models** (2407)
1810 **Bonnie Doll** cut-out book, #2751 for doll & #2750 Elizabeth for clothes
1810 **Boy and Girl - Teen Dolls** (2407)
1811 **Cut-Out Book - Betsy Doll** (3903)
1811 **Cutie Doll** cut-out book (1283)
1812 **Cut-Out Book Susan Doll** (3903)
1812 **Hollywood Glamour** (2407)
1813 **Cut-Out Book - Sally Doll** (3903)
1813 **Three Little Maids From School Are We** (2404)
1814 **Bride Doll Book** (2493)
1814 **Cut-Out Book - Cindy Doll** (3903)
1815 **Going Places With Mary and Bill** (2563)
1815 **Nancy and Her Dolls** (2750 Elizabeth) some clothes taken from #2751 Bonnie
1816 **Baby to Dress and Care For** (2749 Baby Anne)
1817 **Mimi** (2422)
1818 **Emily** (2423)
1819 **Trudy** (2424)
1820 **Vicky** (2762)
1821 **Julie** (2763 Dolly Gets New Clothes)
1822 **Cut-Out Book Susan** (3903)
1823 **Cut-Out Book Mary** (1829 Playmates)
1824 **Cut-Out Book Sally** (3903)
1825 **Cut-Out Book Cindy** (3903)
1825 **Doll Party, Sue and Penny** (4207)
1826 **Punch-Out and Dress Betsy and Bitsy** (1283)
1826 **Cut-Out Book - Betsy Doll** (3903)
1827 **Teen Age Dolls** (1253 Prom)
1827 **Cut-Out Book - Betty Doll** (1831 Lollypop Kids)
1828 **Mother's Girl and Grandmother's Doll** (2561)
1828 **Dress Dolly Cut-Outs - Lucy** (1886)
1829 **Dress Dolly Cut-Outs - Janie** (1887)
1829 **Playmates** 1961 (date in Lowe records)
1830 **Dress Dolly Cut-Outs - Patty** (1885)
1830 **Fashion Models** (2585)
1831 **Playhouse Dolly "Jo"** (2784 Little Girls)

1831 **Lollipop Kids** 1961 (date in Lowe records)
1832 **Playhouse Dolly "Jill"** (2784 Little Girls)
1832 **Lots of Dolls** 1961 (date in Lowe records) dolls new, clothes from #1057
1833 **Young Couple** (2407)
1833 **Playhouse Dolly - Jane** (2784 Little Girls)
1834 **Baby Doll** (2749 Baby Anne)
1834 **Honeymoon Dolls** (2493)
1835 **Doll Book - Sally and Sue** (2785)
1835 **Twins - Jane and Jill** (4219)
1836 **Dolls - Molly and Nina** (1832 Lots of Dolls)
1836 **Mary Jane** - 10" Doll Punch & Dress (1283)
1837 **Brother and Sister Dolls** (1832 Lots of Dolls)
1837 **Fashion Models - Laura and Joanne** (2407)
1838 **Paper Doll Playmates - Tim and Barbara** (1832 Lots of Dolls)
1838 **Teen-Age Dolls** (2483)
1839 **Dress-Up Doll Book - Sherry** (2785)
1839 **Sisters** (2411)
1840 **Margo Travels to Foreign Countries** (2714)
1840 **Baby Sitter** (1025 Turnabout & 1946 Clothes Crazy)
1841 **Jill - Doll Cut-Out Book** (2784 Little Girls)
1842 **Jane - Doll Cut-Out Book** (2784 Little Girls)
1842 **A Box of 10 Cut-Out Dolls** 1947 (date in Lowe records - John Martin's House trademark)
1842 **Cuddly Cut-Outs** (1252 Rockabye)
1843 **Cover Girls** (2407)
1843 **Sandra - Doll Cut-Out Book** (2717 Grow Up)
1844 **Doll Time** (1284 & one small doll from 9041)
1844 **Janet - Doll Cut-Out Book** (2717 Grow Up)
1845 **Shirley - Doll Cut-Out Book** (2717 Grow Up)
L1845 **Two Story Doll House with Paper Dolls** from #1028 Playhouse, made for Sears
1845 **Teen-Age Sister Dolls** (2411)
1846 **Doris - Doll Cut-Out Book** (2721)
1846 **The Girls Next Door** (2562)
1847 **Playhouse Paper Dolls** box of six paper dolls from the 1940's; books varied as to titles available in stock
1847 **Sherry and Terry** (2722 Little Dolls)
1847 **Sisters** (2411)
1848 **Sue - Doll Cut-Out Book** (2721)
1849 **Sally - Doll Cut-Out Book** (2785)
1850 **Jenny and Nancy - Doll Cut-Out Book** (2723)
1851 **Peggy and Trudy - Doll Cut-Out Book** (2723)
1851 **All American Girls** box, combined dolls from four books: #1043 Bride Doll, #1048 The Turnabouts, #1046 Clothes Crazy & #1026 Dude Ranch
1852 **Playtime Pals** large envelope contained assorted paper doll books from the late 1940's & early 1950's; books varied as to titles available in stock
1854 **City Girl** 1963 (3903)
1855 **Mammoth Box of Dolls and Dresses** 29 dolls from assorted books of the 1940's
1855 **Schoolgirl Doll Book** 1963 (3903)
1856 **Look Alike - Dress Alike** 1963, Twins Doll Book (2404)
1857 **Vacation** 1963 (4207)
1858 **What Shall We Wear** 1963 (2404)
1860 **Mother and Daughter** 1963 (2411)
1861 **Nurse Cut-Out Doll** 1963 (4913)
1862 **Masquerade Party** 1963 (1832 Lots of Dolls)
1863 **Doll Book - Sherry and Jerry** 1963 (2407)
1864 **Society Dolls** 1963 (2488)
1865 **Dolly Goes Shopping** 1964 (3921 Patti)
1866 **Baby Dolls** (1252 Rockabye)
1867 **Schoolmates** 1964 (2743 Two Teens)
1868 **Sister Dolls** 1964 (2407)
1869 **Debbie and Diane** 1964 (2562)
1870 **Nurse Cut-Out Doll Book** 1964 (4913)
1871 **City Girl** 1964 (3903)
1872 **Schoolgirl** 1964 (3903)
1873 **Rosemary Clooney** box (   )
1873 **What Shall We Wear** (2404)
1874 **Penny and Sue** 1964 (4207)
1875 **Susan - Clothes For School** (2743 Two Teens)
1876 **Playmates Doll Book** 1964, Mary and Bill (1056)
1877 **Hello, Patti** 1964 (3921)

1878 **Hair-Do Doll** 1964 (2743 *Two Teens*)
1879 **Sister Dolls** 1964 (2411)
1880 **Airline Stewardess** 1965 (4913)
1881 **Penny Goes to School** 1965 (3903)
1882 **Baby Dolls to Dress and Care For** 1965 (1252 *Rockabye*)
1883 **Dolly Goes Shopping** 1965 (3903)
1884 **Schoolgirl** 1965 (2404)
1885 **Patty - Little Dressmaker Doll Book** 1966
1886 **Lucy - Little Dressmaker Doll Book** 1966
1887 **Janie - Little Dressmaker Doll Book** 1966
1889 **Susan - Hair-Do-Doll** (2743 *Two Teens*)
1890 **Jackie - Hair-Do-Doll** (2743 *Two Teens*)
1891 **Bridesmaid Dolls** (2562)
1892 **Mary** (1829 *Playmates*)
1893 **Betty** (1831 *Lollipop Kids*)
1894 **Sally** (1831 *Lollipop Kids*)
1902 **Box of Dolls** ( )
1915 **Sew on Dolls** (2610)
1919 **Little Girl Dolls 'N Dresses** (4207)
1941 **Let's Play House** box, six books that may include any of the following: #969 **Hollywood Dolls**, #946 **Down on the Farm**, #1284 **Toni**, #945 **Baby Sitter**, #968 **Square Dance**, #1248 **Let's Play House**, #1327 **Lollypop Crowd**, #958 **Career Girls**, & there may be others
2103 **Toytown Motorama** 1962 catalog, stand-ups
2104 **Cowboys and Indians** 1961, stand-ups
2105 **Play Housekeeping** 1961 (1248)
2107 **Model Village** 1961, stand-ups
2108 **Punch-Out Bazaar** 1962 catalog, toys & stand-ups
2109 **Mask-A-Rade** 1962 catalog, masks
2109 **Maskarade** 1969 masks
2109 **Maskarade** 1973 masks (2109)
2110 **Santa's Band** 1962, has two paper dolls & stand-ups
2110 **Santa's Christmas Punch-Out Book** (2110 *Santa's Band*)
2115 **Punch-Out Circus** 1963, stand-ups
2116 **Toytown Motorama** 1963, stand-ups
2117 **Punch-Out Toys For Girls and Boys** 1963, stand-ups by John Dukes McKee
2118 **Zoo** 1966, stand-ups
2118 **Zoo** 1968, stand-ups
2118 **Zoo** 1971, stand-ups
2120 **Toy Town** 1966, stand-ups
2122 **Flag Bingo** 1966, game
2124 **Picture Dominoes** 1966, game
2125 **Punch and Judy Show** 1967 & 1971, puppets, stage & script
2127 **Doll House Furniture** stand-ups
2128 **Play City** 1967, stand-ups
2129 **Circus** 1967, stand-ups
2131 **Fire Engine** 1967, stand-ups
2132 **Farm Animals** 1969 catalog, stand-ups
2134 **Toyland** 1970, paper toys
2135 **Punch-Out and Play Toys for Boys** 1970, punch-out toys
2136 **Fire Department** 1970, punch-out book
2137 **Circus Punch-Out** 1970, stand-ups
2138 **Punch-Out Finger Puppets** 1971
2139 **Playtown Punch-Out** 1971, stand-ups
2140 **Stencil Punch-Out Sign Shop** 1971
2141 **Stencils** 1972
2211 **Fire Department** 1962, punch-out & coloring book
2212 **Build a City** 1962, coloring book & punch-outs
2213 **Mother Goose Toy Book** 1962, coloring book & stand-ups
2216 **Main St.** 1963, coloring book & stand-ups
2217 **Indians** 1963, coloring book & stand-ups
2218 **Real Cowboys** 1963, coloring book & stand-ups
2221 **Planes** 1965, coloring book & model planes
2225 **Punch-Outs and Coloring** 1964
2239 **Party Fun** 1968, coloring & decorations
2240 **Puppet Punch-Out Marionettes** 1969 catalog, coloring book & punch-outs
2241 **Cowboys** 1969 catalog, coloring book & stand-ups

2242 **Indians** 1969 catalog, coloring book & stand-ups
2360 **His Peek-A-Book** 1961, story book with costume on each page (2563)
2361 **Her Peek-A-Book** 1961, story book with costume on each page (2563)
2403 **Baby Doll** 1957
2404 **Three Little Maids From School Are We** 1957
2405 **Janet Leigh** 1957
2406 **Patti Page** 1957
2407 **Bob Cummings Fashion Models** 1957
2408 **The Three Cheers** 1957 (1253 *Prom*)
2411 **Patience and Prudence** 1957
2419 There are over 30 different packets with this number. Each consists of paper doll books with other numbers mentioned elsewhere in this list. Usually two books are in each packet but not always.
2420 **Six Cut-Outs and Lovely Clothes** (1832 *Lots of Dolls*)
2421-A **Alive-Like Baby Doll** (2749 *Baby Anne*)
2421-B **Alive-Like Dolly** (2751 *Bonnie*)
2421-C **Alive-Like Dolly** (2750 *Elizabeth*)
2422 **Mimi - Doll Keepsake** 1964
2423 **Emily - Doll Keepsake** 1964
2424 **Trudy - Doll Keepsake** 1964
2426 **Fashion Designer - Patty** (1885)
2426 **Fashion Designer - Lucy** (1886)
2426 **Fashion Designer - Janie** (1887)
2427 **Little Miss Paint by Numbers** - Dress Designer Set (2574 *Betsy*)
2431E **Six Cut-Out Paper Dolls** (1832 *Lots of Dolls*)
2438-A **Little Miss Sew and Sew** - Clothes Lace-On (1283)
2438-B **Little Miss Sew and Sew** - Clothes Lace-On (1283)
2438-C **Little Miss Sew and Sew** - Clothes Lace-On (1283)
2480 **Rosemary Clooney** 1958 (2585)
2481 **Janet Leigh - Hollywood Glamour** 1958 (2405)
2482 **Sweetie Pie** 1958 (1283)
2483 **Junior Misses** 1958
2484 **Vacation Album** 1958 (4913)
2485 **Penny's Party** 1958 (4207)
2486 **Million Dollar Dolls** 1958 (2407)
2487 **Rosemary Clooney** 1958 (2585)
2488 **Patti Page** 1958
2489 **Rosemary Clooney** 1958 ( )
2490 **Brother and Sister** 1959 (2563)
2491 **School Days** 1959 (2404)
2492 **Sweetest Baby** 1959 (2403)
2493 **Bride and Groom** 1959
2494 **Sisters** 1959 (2411)
2502 **Christmas Fun Book** 1974, coloring book & paper toys
2508 **Pam and Her Pram** 1956, paper doll & coloring book, paper dolls from #1057, #1284, #2574 *Betsy*
2515 **Gloria's Make-Up Paper Doll** (2585)
2522 **Paper Playmates, Jan and Jimmy** 1955 (2563)
2523 **Kay and Kim with Clothes to Trim** 1956
2525 **Happy Days Fun Book** 1961, includes paper dolls
2528 **Annie the Action Doll** 1954 (1283)
2530 **Little Miss Muffet** (1283)
2530 **Rosemary Clooney** coloring book with paper dolls (2585)
2531 **Betsy and Bob** (1286)
2532 **Patience and Prudence** 1957, coloring book, includes paper dolls (2411)
2539 **Children's Party Pack** contains three assorted paper doll books
2553 **Merry Christmas Activity Book** 1956, includes a paper doll
2556 **Sewing Card Book with Sew On Clothes** 1953, five dolls from three different books: #1045 *Playtime Pals*, #1284 *Toni* & #1252 *Rockabye*
2559 **Dotty Dimple** (1283)

2560 **The Honeymooners** 1956
2561 **Goldilocks and the Three Bears** 1955
2562 **Here Comes the Bride** 1955
2563 **His and Hers** The Guest Towel Dolls 1955
2564 **Teen Queens** 1955 (1253 *Prom*)
2565 **Schoolmates** (1049 *Lollypop*)
2566 **Rosemary Clooney** 1954 (2585)
2567 **Kit and Kapoodle** 1954 (4267 & 4268)
2568 **Betty Buttercup** 1956 (1283)
2569 **Rosemary Clooney** 1956
2572 **Tanks and Planes** 1943, stand-ups
2573 **Soldiers and Sailors** 1943, stand-ups
2574 **Doll House Furniture** 1943, stand-ups
2574 **Paint Betsy's Sunday Best** 1955 (date in Lowe records) paint book, doll & clothes to be painted, brush & paints included
2575 **Barnyard Animals** 1943, stand-ups
2576 **Ten Dolls** box, 1947 catalog (1025 *Turnabout Dolls*)
2582 **Annie the Action Doll** 1954 (1283)
2582 **Happy Birthday To You** 1954, activity book with paper dolls
2585 **Gloria's Make-Up** 1952 (date in Lowe records) This was the original of the Rosemary Clooney paper dolls #1256 & #2713 and their reprints. The dolls were redrawn & six of the eight pages of clothes used in the Gloria Book were retained in the Rosemary Clooney books.
2588 **Do It Activity Book** 1955, one page paper doll from #1057 & #1028 *Playhouse*
2589 **Five Wood Dolls** box (1025 *Judy*)
2589 **Every Which Way** 1956, activity book with a paper doll
2591 **Six Pressboard Dolls** 1942, box (1021 *Baby Show*)
2593 **High Chair to High School** 1952 catalog, box, cut-out dolls. Contained five paper doll books of assorted titles from Lowe stock.
2595 **Rosemary Clooney Paper Doll and Coloring Book** 1956 (2585)
2596 **Two in One Color and Cut-Out Book** paper dolls from #4207 and #1057
2601 **Rosemary Clooney** coloring book with paper doll, some books just have clothes and no doll
2602 **Rosemary Clooney** coloring book with paper doll, some books just have clothes and no doll (2585)
2610 **Sally's Dress a Doll Storybook** 1952 (date in Lowe records) Some of these dolls were produced with a child's handkerchief for each day of the week.
2612 **TV Coloring Book** includes paper doll of Rosemary Clooney & little girl Sally to color.
2613 **School Days Coloring Book** includes paper dolls
2639 **Playtime** coloring book with paper dolls from #1283 & #1252 *Rockabye*
2701 **Annie the Statuette Doll** 1956 (1283)
2702 **Party Dolls** 1956 (990)
2703 **Fashion Models** 1956 (1048 *The Turnabouts*)
2704 **Betsy and Bob** 1956 (1286)
2705 **Kit and Kapoodle** 1956 (4267 & 4268)
2706 **Dolly and Me** 1956 (1284)
2707 **Square Dance** 1957 (968)
2708 **Let's Play House** 1957 (1248)
2709 **Fairy Princess** 1957 (1242)
2709 **Dollies Go Shopping** (2422 & 2423)
2710 **Dolls and Dresses - Patty and Jan** 1971 (1885 & 1887)
2710 **Teen Queens** 1957 (1253 *Prom*)
2711 **Dotty Dimple** 1957 (1283)
2711 **Dolls and Clothes** 1971 (2422 & 2423)
2712 **Two Teens** 1971 (2743)
2713 **Rosemary Clooney** 1957, originated from #2585 *Gloria's Make-Up*; six of eight pages of clothes were retained
2713 **Mopsy and Popsy** 1971
2713 **2 Cut-Out Doll Books** in package with scissors. Books are #1513 **Mother & Daughter** & #1514 **Ruth, Joan & Winnie**. The books may vary in the packages.
2714 **Dollies Go 'Round the World** 1971

2715 **Dress the Dollies - Penny and Sue** 1971 (4207)
2716 **Country and Western** 1971 (968)
2717 ~~**Sew On Doll and Clothes** - Little Miss Sew and Sew (1283)~~
2717 **When We Grow Up** 1971
2718 **Here Comes the Bride** 1971
2719 **Mother and Daughter** 1971 (2411)
2719 **I'm a Big Cry Baby** doll has noise maker (2403)
2720 **Dress Annabelle** 1972
2721 **Doll Friends** 1972
2722 **Little Dolls** 1972
2722 10" statuette doll with five dresses to sew on with yarn (1283)
2723 **A House Full of Dolls and Clothes** 1972
2724 **Now and Then Paper Dolls** 1973
2725 **The Happy Family** 1973
2728 **Dress Up Dolls** 1969 (2411)
2730 **Polly Pal** (3730)
2730 **Polly and Her Dolly** 1971 (9118)
2731 **Rosemary Clooney** 1958 (2585)
2731 **Schoolmates** (1049 Lollypop)
2732 **Bob Cummings Fashion Models** 1958 (2407)
2732 **Fashion Models** (1251 Toni)
2733 **Janet Leigh** 1958 (2405)
2733 **Career Girls** (958)
2734 **Cinderella** 1960 (1242)
2734 **Patti Page** 1958 (2406)
2735 **Glamorous High Fashion** 1958 (2407)
2735 **After School Party - Bandstand** (2483)
2736 **Duet Dolls** featuring Patience & Prudence, 1958 (2411)
2736 **His and Hers** - The Guest Towel Dolls (2563)
2737 **Here Comes the Bride** 1962 (2493)
2738 **Schoolgirls** 1962 (2404)
2738 **Cuddles and Rags** (1283)
2739 **Patti Page** 1959 (2406)
2739 **Airline Stewardess** 1962 (4913)
2740 **Dr. Kildare and Nurse Susan** 1962, Susan dolls came from #4913 Airline Stewardess with some clothes. Dolls were redrawn. Kildare doll came from #2407 Bob Cummings with some clothes. Doll was redrawn. There are two different front covers: blue background & signed by Pollard, & unsigned red background
2740 **Teens** 1959 (2483)
2741 **Cheerleaders** 1962 (2483)
2741 **Party Dolls** 1959 (4207)
2742 **Twinkles** 1966 (2563)
2742 **School Children** 1963 (1832 Lots of Dolls)
2742 **Airline Stewardess** 1959 (4913)
2743 **Mimi** 1966, new clothes, doll from #3905
2743 **Celebrity Fashion Show** (2407)
2743 **Two Teens** 1963
2744 **Baby Sister** 1966 (2403)
2744 **Little Miss Muffet** 1959 (1283)
2745 **Dress Up Dolls** 1963 (2560)
2745 **My Big Dolly** 1966 (5908)
2745 **Cuddles** (5908 Molly Dolly)
2748 **Celebrity Fashion Show** 1959 (2407)
2748 **Saturday Night Barn Dance** (968)
2749 **Dolls and Their Dollies** 1961 (1284)
2749 **Baby Anne Smiles and Cries** 1964
2750 **Alive Like Doll - Elizabeth** 1963
2750 **College Girls** (1046 Clothes Crazy)
2750 **Royal Princess** 1961 (2406 & 2488)
2751 **Bonnie - Alive Like Dolly** 1963
2751 **Big Doll Betty** 1960 (date in Lowe records)
2752 **School Time** 1964 (1832 Lots of Dolls)
2752 **Look Alike Twins** (2904)
2753 **Dollies Go Shopping** ©1964 & 1969 (2422 & 2423)
2753 **Six School Girl Dolls** 1961 (2404)
2754 **The Jones Family** 1961 (2562)
2754 **Betty Buttercup** 1964 (1283)
2754 **The Bobbsey Twins** (1254)
2755 **Bride and Groom** 1961 (2493)
2755 **Sally and Jane** 1964 (2743 Two Teens)
2756 **Life Like Animated Faces Doll** 1964 (2751 Bonnie)
2756 **Rock-A-Bye Babies** (1252)

2757 **Molly Dolly** 1965 (5908)
2757 **Little Calendar Girl** 1961, doll's head from #2906, rest of doll & clothes from #2751 Betty
2757 ~~**Janet Leigh** 1957 (2405)~~
2758 **Penny's Party** 1965 (4207)
2758 **Patti Page** 1957 (2406)
2758 **Little Ballerina** 1961 (2915)
2759 **Fifth Avenue Doll Book** 1961 (2405)
2759 **Patti Alive-Like Doll Book** 1965 (3921)
2759 **Real Nurse** (958 Career Girls)
2760 **Dotty Doll Book** 1968, doll is redrawn, clothes from #2743 Mimi
2760 **Dolls** 1961 (1046 Clothes Crazy)
2761 **Dollies Go To School** (2404)
2761 **Baby Doll** 1957 (2403)
2761 **Sister Dolls** 1964, also dated 1966 (2411)
2761 **My Very Own Baby Doll** 1967 (2749 Baby Anne)
2762 **Three Young Ladies** 1964 (2404)
2762 **Girls Department** (2404)
2762 **Little Maids Dress Shop** 1957 (2404)
2762 **Vicky** 1967 (#2751 Bonnie)
2763 **Dolly Gets Lots of New Clothes** 1967 (2750 Elizabeth)
2763 **Hootenanny - Country and Western** 1964 (968)
2764 **Pixie Doll and Pup** 1968
2764 **Dress Up Animals** (4267 & 4268)
2764 **Archies Girls** 1964 (2483)
2765 **Western Cowgirl and Cowboy** (1286)
2765 **We're the Jones Family** 1958 (2562)
2766 **Front and Back Dolls and Dresses** 1964
2766 **Cut-Out Dolls Western Style Clothes** 1958 (968)
2766 **Hair-Dos** 1960 (2493)
2767 **Dimples** 1958, Best Dressed Baby Doll in Town (2403)
2767 **Goldilocks** 1964 (2561)
2768 **Tina and Her Friends** 1960 (1253 Prom)
2768 **Campus Queens** (1253 Prom)
2768 **Dr. Kildare** nurse dolls from #4913 & Dr. from #2407, dolls redrawn
2768 **Hello Dollies** 1964 (2424 for large doll & 9118 for small doll)
2769 **Schoolmates** 1960 (1056)
2769 **Career Girls** (958)
2769 **Dollies Go To School** 1971 (2404)
2770 **Holiday Cruise** 1965 (2483)
2770 **Cinderella** 1958 (1242)
2771 **Suzy** 1966 (5901)
2771 **Hollywood Glamour - Janet Leigh** 1958 (2405)
2772 **After School Party - Bandstand** 1958 (2483)
2772 **Hello Patti** 1967 (3921)
2773 **Schoolmates** 1958 (2404)
2773 **Schoolmates** 1967 (2743 Two Teens)
2774 **Rosemary Clooney** 1958 (2569)
2774 **Twinkles** 1967 (2563)
2775 **Polly and Her Dolly** 1968 (9118)
2775 **Boy Meets Girl - Jerry and Sherry** 1958 (2407)
2775 **Sally and Jane** 1964 (2743 Two Teens)
2776 **Molly Dolly** 1968 (5908)
2776 **Fabulous High Fashion Models** 1958 (2407)
2777 **Betty Doll** 1968 (2563) dolls & clothes redrawn
2778 **Dress Up Dolls** 1969 (2411)
2779 **Dolls and Dresses** 1969 (1885, 1886, 1887)
2780 **Dollies Try On New Clothes** 1969 (2762 Vicky & 2763 Dolly)
2781 **Girls and Boys** dated 1969 & 1973 (1832 Lots of Dolls)
2782 **Dollies Fashion Show** 1969 (2750 Elizabeth & 2751 Bonnie)
2782 **Girls Department** (2404)
2783 **Three Models and Hair-Dos** (2493)
2783 **Dolltime Cut-Out Book** 1969 (1829 Playmates & 1831 Lollypop Kids)
2784 **Little Girls** 1969
2784 **Hair-Do Dolls** (2411)
2785 **Brenda Lee** 1961 (2407 Bob Cummings) dolls redrawn, some new clothes added
2785 **Gloria's Make-Up** (2585)

2785 **Sally, Sue and Sherry** 1969
2786 **Dr. Kildare** 1962, nurse dolls from #4913 & Dr. from #2407, dolls redrawn
2786 ~~**Baby Sue** 1969 (2403)~~
2787 **Fairy Tale Princess** 1962 (1242)
2787 **Little Miss Muffet** 1969 (1283)
2789 **Fashion Models** (1251 Toni)
2791 **Down on the Farm** (1056)
2792 **Girl Friends** (2411)
2793 **Here Comes the Bride** (2493)
2794 **Storybook Dolls** 1959 (2563)
2795 **Goldilocks** 1959 (2561)
2796 **Betty Buttercup** (1283)
2797 **Busy Teens** 1959 (2483)
2802 **Statuette Dolls** (1049 Lollypop Crowd)
2861 **Jolly Santa** 1965 coloring book with Brownie paper doll
2873 **Duet Dolls** paper doll/coloring book, dolls from #2411
2886 **Baby Sue** 1969 (2403)
2888 **Indoor Fun Book** 1968, includes paper dolls
2893 **Punch-Out Toys** 1961, stand-ups
2903 **11" Doll with Button-On Clothes** (2563)
2904 **Sew-On Sherry Ann's Clothes** 1960
2906 **Betsy Dress A Doll Storybook** 1960
2915 **Wendy Dress A Doll Storybook** 1959, new doll, clothes same as #2610
2917 **Four Statuette Dolls** (Dolls can vary)
2919 **Patti Page** coloring book w/paper dolls from #2406
2926 **Patience & Prudence Coloring Book** 1957, includes paper dolls
2930-2 **Sister Dolls** (2411)
2930-1 **Here Comes the Bride** 1962 (2493)
2930-2 **Schoolmates** 1962 (2404)
2930-3 **Girl Friends** 1962 (2411)
2930-4 **Dimples** 1962, Best Dressed Baby Doll in Town (2403)
2932 **Look I'm Alive-Like** (3903)
2933 **Punch-Out Book of Miniature Airplanes** 1954
2955 **Coloring Book Just for Girls** with paper dolls from #1057, #2411 & #2574
2990 **Sally's Sew On Clothes** (2610)
2992 **Make Believe Beauty Parlor** 1959, wigs to punch-out & put-on life-size head
3039 **Playland Fun Book** 1965, includes paper dolls
3082 **Round-up of Things to Color** 1963, includes paper dolls
3092 **Dr. Kildare Play Book** circa1963, includes paper dolls
3362 **Masquerade Party Peek-a-Book** (4299)
3363 **Sherlock Bones Peek-a-Book** (4263)
3364 **Treasure Land Peek-a-Book** (4298)
3365 **Dolly Goes Around the World Peek-a-Book** (4264)
3366 **Shadow Peek-a-Book** (4267)
3367 **Kitty Peek-a-Book** (4268)
3368 **Alfie Peek-a-Book** (4344)
3379 **His Peek-a-Book** (2563)
3380 **Her Peek-a-Book** (2563)
3711 **Little Dolls** 1974 (2722)
3711 **Four Push-Out Dolls to Dress** 1976 (2722 Little Dolls)
3712 **Ms. Dolls** 1976 (2717 Grow Up)
3712 **When We Grow Up** 1974 (2717)
3713 **Doll Push-Outs** 1974 (2721)
3714 **Doll Push-Outs** 1974 (2785 Sally, Sue and Sherry)
3715 **Now and Then** 1976 (2724)
3716 **Baby Sister** 1974 (2403)
3717 **Mopsy and Popsy** dated 1974 & 1976 (2713)
3718 **Goldilocks and the Three Bears** 1975 (2561)
3719 **My Very Own Baby Doll** 1974 (2749 Baby Anne)
3720 **Dollies Go Round the World** 1974 (2714)
3721 **Two Teens** 1974 (2743)
3722 **Little Girls** 1974 (2722)
3723 **Here Comes the Bride** 1975 (2718)
3724 **A House Full of Dolls and Clothes** 1975 (2723)
3725 **The Happy Family** 1973 (2725)

3726 **School Children** 1975 (1832 Lots of Dolls)
3727 **Jeannie and Gene** published in 1975 & 1976
3730 **Polly Pal** 1976
3851 **Punch-Out Merry-Go-Round** 1974, stand-ups
3852 **Picture Dominoes** 1974, game (2124)
3853 **Old Maid & Fortune Telling Cards** 1974
3854 **Flag Bingo** 1974, game (2122)
3855 **Dollhouse Furniture** 1967, stand-ups
3856 **Punch-Out and Fold - Play City** downtown, 1967, stand-ups (2128)
3903 **Mary Ann** animated face, 1960 (date in Lowe records)
3903 **Sally Ann** animated face, 1960 (date in Lowe records)
3903 **Betty Ann** animated face, 1960 (date in Lowe records)
3905 **Mimi From Paree** 1960
3906 **Betsy Doll and Clothes** (2906)
3909A - **Stand-Up Doll-Sew On** (2760 Dotty)
3909B - **Stand-Up Doll-Sew On** (2563)
3909C - **Stand-Up Doll-Sew On** (2563)
3909D - **Stand-Up Doll-Sew On** (2720)
3909E - **Stand-Up Doll-Sew On** (2720)
3909F - **Stand-Up Doll-Sew On** (2720)
3910 **Dress Maker Set** six different sets, four dolls from #1831 Lollipop Kids (two are redrawn), two from #1829 Playmates (one redrawn).
3911 **Make Dolly's Clothes** Fashion Designer Set - Janie (1887). Three different sets: Patty (1885), Lucy (1886) & Janie (1887).
3912 **My Big Dolly To Dress and Read To** (5908 Molly Dolly)
3920 **Trixie** 1961 (2403)
3921 **Playmates - Doll House Dollies** 1966 (1885, 1886, 1887)
3921 **Patti Doll Book** 1961, doll new, clothes from #3903
3922 **Doll House Dollies** 1966 (1829 Playmates & 1831 Lollipop Kids)
3923 **Schooldays - Doll House Dollies** 1966 (1832 Lots of Dolls)
3924 **Sisters - Doll House Dollies** 1966 (2411)
3925 **Doll House Babies** 1966 (1252 Rockabye)
3926 **Goldilocks - Doll House Dollies** 1966 (2561)
3937 **Eyes on Margie - Dress A Doll Storybook** (2751 Betty)
3938 **Eyes on Nancy - Dress A Doll Storybook** (2904)
3941 **My Dolly Twinkle** (5908)
3942 **My Dolly Kisses** (5908)
3944 **Nancy Dress A Doll Storybook** 1964, doll from #9045, clothes from #2904
3945 **Wendy - Storybook Doll** 1964, doll from #9045, clothes from #2915
3946 **Betsy - Storybook Doll** 1964, doll from #9118, clothes #2906
3947 **Cindy - Storybook Doll** 1964, all original
3967 **Little Sister** 1969 ( )
3972 **Fairground** (1049 Lollypop)
4008 **My First Paper Doll Book** Bonnie Book (1057)
4009 **Bye Baby Bunting** 1954, Bonnie Book (4219)
4171 **Gabby Hayes** 1954, Jack-in-the-Book Bonnie Book
4206 **Me and Mimi** 1952, also 1953 Bonnie Book (1283)
4207 **Penny's Party** 1952, Bonnie Book
4208 **My Very First Paper Doll Book** 1952, Bonnie Book (1057)
4214 **Penny and Sue** 1953, Bonnie Book (4207)
4219 **Bye Baby Bunting** 1952, also 1953 Bonnie Book
4220A **Polly Dolly** 1969 (9118)
4220B **Molly Dolly** 1969 (2764 Pixie)
4220C **Baby Doll** 1969 (2403)
4238 **Dolly Takes a Trip** 1957, Bonnie Book (4283)
4263 **Sherlock Bones** 1955, Bonnie Book
4264 **Dolly Goes Round the World** 1955, Bonnie Book
4265 **Billy Boy** 1953, also 1955, Jack-in-the-Book Bonnie Book
4266 **Betty Plays Lady** 1953, Jack-in-the-Book Bonnie Book

4267 **Popsy** 1953, Jack-in-the-Book Bonnie Book
4268 **Trinket** 1953, Jack-in-the-Book Bonnie Book
4270 **Mimi** 1953, Bonnie Book (1283)
4281 **Circus Time** 1952, Jack-in-the-Book Bonnie Book
4282 **Little Sugar Bear** 1952, Jack-in-the-Book Bonnie Book
4283 **Dolly Takes a Trip** 1952, Jack-in-the-Book Bonnie Book
4284 **Cookie the Rabbit** 1952, Jack-in-the-Book Bonnie Book
4298 **Jungletown Jamboree** 1955, Bonnie Book
4299 **Masquerade Party** 1955 Bonnie Book
4308 **My Favorite Doll Book** 1954, Bonnie Book (1057)
4343 **Captain Big Bill** 1956, Jack-in-the-Book Bonnie Book
4344 **Alfie** 1956, Jack-in-the-Book Bonnie Book
4345 **Timmy** 1956, Jack-in-the-Book Bonnie Book
4420 **Penny's Party** Bonnie Book (4207)
4501 **Tall Tales** 1963, Jack-in-the-Book Bonnie Book
4501 **I'm Amy - I'm Jill** 1974 (2760)
4502 **Circus** 1964 Jack-in-the-Book Story, listed in 1964 Lowe catalog
4502 **I'm Mimi - I'm Judy** 1974 (2720)
4503 **Big Bill** 1963, Jack-in-the-Book Story (4343)
4503 **I'm Ginny - I'm Lolly** 1974 (2563)
4504 **I'm Trudy - I'm Katie** 1974 (1283)
4505 **School Children** 1975, also 1978 (1832 Lots of Dolls)
4506 **Sally and Jane** 1964, also 1978 (2743 Two Teens)
4507 **A House Full of Dolls** 1975 (2723)
4508 **Ms. Dolls** 1976 (2717 Grow Up)
4509 **Now and Then** 1976 (2724)
4509-D **Sturdy Stand-Up Dolly** lace-on (2720)
4509-E **Sturdy Stand-Up Dolly** lace-on (2720)
4509-F **Sturdy Stand-Up Dolly** lace-on (2720)
4510 **Jeannie and Jean** 1976 (3727)
4531 **Here Comes the Bride** (2718)
4532 **The Happy Family** 1973 (2725)
4533 **Dollies Go Round the World** (2714)
4534 **Polly Pal** (3730)
4535 **Goldilocks and the Three Bears** 1975 (2561)
4536 **Little Girls** 1974 (2784)
4601 **Alive-Like Animated Doll** (1242)
4701 **Little Boy Meets Little Girl** 1957, Bonnie Book (1254)
4704 **Adventures of TV Tim** 1957, Bonnie Book, new doll, rest of book like #4171
4730 **Bye Baby Bunting** 1957, Bonnie Book type (4219)
4731 **Me and Mimi** 1957, Bonnie Book (1283)
4732 **My Very First Paper Doll** Bonnie Book (1057)
4851 **Fire Department** 1975, stand-ups
4852 **Zoo Stand-Up Animals** 1975, stand-ups (2118)
4853 **Punch-Out and Play Toys For Boys** 1975, paper toys
4854 **Punch and Judy** 1975, puppets (2125)
4856 **Finger Puppets** 1975, puppets
4857 **Punch-Out Shapes** 1976
4858 **Farm Animals** 1976 stand-ups
4859 **Mask-A-Rade** punch-outs, masks
4881 **Molly Dolly** 1969 (5908)
4882 **Goldilocks and the Three Bears** 1969 (2561)
4883 **Patti, Pink - Pert and Petite** 1969 (3921)
4884 **Suzy** 1969 (5901)
4886 **School Boy Doll** 1963 (9301)
4887 **School Girl Doll** 1963 (9302)
4909-D **Stand-Up Dolly** 1972, lace-on clothes (2720)
4909-E **Stand-Up Dolly** 1972, lace-on clothes (2720)
4909-F **Stand-Up Dolly** 1972, lace-on clothes (2720)
4913 **Airline Stewardess** 1957
4930-A **Wendy** 1970, doll from #9045, clothes from #2915
4930-B **Nancy** 1970, doll from #9045, clothes from #2904
4930-C **Cindy** 1970 (3947)
5003 **A Playtime Box of Paper Dolls** circa 1944, contained an assortment of paper-doll books

which could vary according to available titles
5751 **Giant Fun Pack** in 1966 catalog, contents appear to be same as #9211
5752 **World Wide Activity Pack** in 1966 catalog, contents appear to be same as #9212
5753 **Betsy Doll Pack** (9118)
5754 **Doll Baby Pack** (2403)
5755 **Dressmaker Set** (2750 Elizabeth & 2751 Bonnie)
5756 **Hello Dollies Doll Pack** (3947 & 1832 Lots of Dolls)
5759 **Keepsake Dolls Read and Dress** 4 different packets with this number, each contained one of the Dress-A-Doll Storybooks, either #3944, #3945, #3946 or #3947 plus 2 small paper doll books which varied
5801 **Patti Page** 1958 (2406) Stauette Dolls in a folder shaped like a purse with make-up kit, make-up chart & appointment book
5840 **Farm Animals** stand-ups
5841 **Punch-Outs of American West**
5842 **Planes - Miniature Models** 1965, stand-ups
5844 **Mother Goose Punch-Out Activity** 1961, stand-ups
5846 **Circus Punch-Out** stand-ups, 1970
5901 **Suzy - Alive Like Face Doll** 1961
5908 **Molly Dolly** 1962, new doll, some clothes from #3905
5908 **My Dolly Twinkle** 1962 (5908 Molly Dolly)
5908 **My Dolly Kisses** 1963 (5908 Molly Dolly)
5909 **Dollies Go Shopping** 1962 (3903)
5909 **Alive-Like Dolls** 1963 (3903)
5909-G **Lace-On Clothes Dolly** 1975 (2760 Dotty)
5909-H **Lace-On Clothes Dolly** 1975 (2760 Dotty)
5909-I **Lace-On Clothes Dolly** 1975 (2563)
5909-J **Lace-On Clothes Dolly** 1975 (2563)
5911-A **Doll Dressmaker** 1972 Janie (1887)
5911-B **Doll Dressmaker** 1972 Patty (1885)
5911-C **Doll Dressmaker** 1972 Lucy (1886)
5961 **Little Sister** 1969 (9041)
5962 **Big Sister Jill** 1969 (9302)
6735 **Goldilocks** 1975 (2561)
6736 **Little Girls** 1974 (2784)
6901 **Fairy Princess Doll** with animated face, 1960 (1242)
6907 **Hello Dollies** (3903)
6918 **Betsy Doll** 1969 (9118)
6919 **Baby Doll** 1969 (2403)
6920 **Make Dollies Clothes** paint by number (2750 Elizabeth & 2751 Bonnie)
7502 **Doll House** 1943, box with doll house & stand-up furniture
8045 **Hello Dollies** 1973, lace-on costumes (2720)
8730 **Keepsake Dolly - Mopsy and Popsy** (2713)
8731 **Keepsake Dolly - Jill and Jane** (2784 Little Girls)
8732 **Dressmaker Doll** (1829 Playmates & 1831 Lollipop Kids)
8733 **Dotty** (2760)
8902 **Lace and Dress Puppy** 1975
8903 **Lace and Dress Kitty** 1975
9000 **Rosemary Clooney** portfolio with paper dolls & coloring book (2569 & 2585)
9041 **Peggy and Peter** 1962, Big, Big, Doll Book
9042 **House For Sale** 1962, dolls & furniture from #1248, no outfits for dolls
9044 **Play City** 1963 catalog, stand-ups
9045 **Two Little Girls** 1964. Dolls new, clothes from #2610, #2915 & #2904
9046 **Punch-Out Bazaar** 1965 catalog, stand-ups & toys
9047 **Brother and Sister** 1965 catalog (9301 & 9302)
9047-P **Brother and Sister** 1967 (9301 & 9302)
9079 **Indoor Fun Activity Book** 1968, two pages of paper dolls to color
9101 **Mimi** clothes snap on (3905)
9102 **Fairy Princess** (1242)
9103 **Three Livin Dolls** (3903)
9104 **Suzy** 1962 box (5901)
9105 **Patti** 1962 box (3921)
9106 **Molly** 1962 box (5908)

WHITMAN PUBLISHING COMPANY, INC.

The Whitman story begins with its parent corporation, Western Publishing Company.

Western Publishing Company began in Racine, Wisconsin as a printing firm founded by Edward H Wadewitz in 1907. Known as the West Side Printing Company, it consisted of five people: Edward H. Wadewitz, Roy A. Spencer, William R. Wadewitz (a brother of Edward), William Bell and Kate Bongarts. The original location was the basement of a millinery store in the 600 block of Racine's State Street.

In 1908 the Company moved one block east to 548 State Street, occupying the first floor and basement. At this time three new presses and a power cutter were added and several more employees joined the payroll.

The business was incorporated in 1909 under the leadership of Roy A. Spencer, the Company's first president. This same year the firm moved from the 548 State Street address just west of the Root River Bridge to the lower floor of the six-story Shoop Building just east of the Bridge. The building was named for a Dr. Shoop whose organization had a widely-developed line of patent medicines sold through the mail. This mail-order operation necessitated a great deal of literature, the printing of which West Side Printing eventually took over.

The Company was organized as a corporation in 1910 and promptly purchased its first lithographic press. This event brought about a name change and the West Side Printing Company became Western Printing and Lithographing Company. When Dr. Shoop retired in 1914, Western acquired his equipment and expanded to all six floors of the building.

A major development occurred in 1916. Western was manufacturing books for the Hamming-Whitman Publishing Co. of 1018 S. Wabash Ave. in Chicago. (Formed in 1912 as the Hamming Publishing Co., its name was changed in 1915 to Hamming-Whitman by the acquisition of Mr. M. A. Whitman's company.) When Hamming-Whitman defaulted on its bills, Western was left with thousands of partly-finished and finished books. Western saw no hope of ever collecting the money due from the Hamming-Whitman Co. As the chief creditor, Western acquired all the assets of the Hamming-Whitman Co. and formed a subsidiary corporation, Whitman Publishing Company.

In the meantime, Western's growth was beginning to saturate the Shoop Building in Racine. Western added a box

The Shoop Building in Racine, WI, an early home of the Western Publishing Co.

department and installed an electrotype department as well. In addition, a working agreement was developed with Racine Engraving Company which established itself on the fifth floor of the Shoop Building. It was only a matter of time before it too was absorbed by Western.

By 1918 the Shoop Building was no longer adequate to house the growing company. A building one block away was acquired and became known as Plant #2. The book binding and box departments were re-located here.

In 1925 (after much encouragement from stores such as Kresge and Woolworth), Western entered into the playing card field. This venture was filled with many problems, particularly in die-cutting the cards to the required extreme accuracy. An engineer by the name of Mr. Hood appeared on the premises and told Mr. E. H. Wadewitz that he could design a machine to cut a certain number of cards a minute with the utmost accuracy. Mr. Hood promised to have the machine in operation in six months time. If it failed, Mr. Hood would not accept a penny in payment. If the machine did all he claimed, Mr. Hood would receive $5,000 cash. Western would get all rights and patents to the machine and Mr. Hood would be on his way with no strings attached. Mr. Wadewitz and his associates agreed to the arrangement.

Days before the six months were up, the machine was completed and performed satisfactorily. Although Mr. Wadewitz tried to persuade Mr. Hood to stay, he declined. The machine he designed was the forerunner of many similar machines used by Western over the years. They are manufactured in the Western machine shops and have been sold to playing card manufacturers all over the world!

In 1927 it was decided to build a new building as the Company by now had overflowed to many different locations. On January 3, 1928, Western moved into its new home on Mound Avenue. This building became known as the Main Plant and was one of the first completely climate-controlled facilities of its kind in the country.

In 1933 Walt Disney characters were introduced to the public through the Big Little Books, coloring books, and other books published by the Whitman Publishing subsidiary. This was the start of a long line of licensed characters and stars of movies, radio, and television which were featured in paper doll books, story books, coloring books, games, and activity sets.

Western decided to establish a facility in the eastern United States in 1934. The plant was located in Poughkeepsie, New York, and many of the employees transferred from Racine to this eastern location. Through the years other plants and offices have been established in Chicago; St. Louis; Los Angeles; Detroit; New York; San Francisco; Washington, D.C.; Hannibal, Mo.; Cambridge, Md; Mt. Morris, Il; Milan, Italy; Paris, France; Weston, Ontario, Canada; and Geneva and Zurich, Switzerland.

A subsidiary known as Artists and Writers Guild, Inc. was established in the 1930's and worked in collaboration with other publishers in planning and producing juvenile books. These offices, located in New York, were later to become known as Artists and Writers Press, Inc.

The Western Newsstand Division collaborated with the Dell Publishing Co. to originate and produce millions of pocket editions and comic books.

In 1960 Western became a publicly owned corporation and adopted a new corporate name: Western Publishing Company. Western Printing and Lithographing Company became the operating unit.

From a very small beginning back in 1907, the Western Publishing Company today ranks among the major corporations in the United States. Its products are manufactured and marketed around the world. Although the Whitman logo was dropped from paper doll books in 1983 in favor of the Golden logo, Western remains a force in paper doll publishing and old Whitman doll books continue to be highly prized by collectors.

The first paper doll book sold by the Whitman Publishing Co. was called **The Dolls That You Love** ($75 - 100) and consisted of six dolls and their clothes. This book had been copyrighted in 1910 by the L. W. Walter Co. It was originally published by the Hamming Publishing Co. of Chicago and later by the Hamming-Whitman Co. This book was then sold by the Whitman Publishing Co. until at least 1918. Whitman later published the six dolls and clothes packaged in an envelope and titled **Cut Out Dolls**. There was no identifying number on either the hard cover book or on the envelope set.

674 Judy Garland Fashion Paint Book 1940 $40 - 60

900 "Our Gang" 1931 $150 - 200

900 Peasant Costumes of Europe 1934 $45 - 65

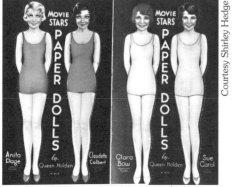

905 Movie Stars Paper Dolls 1931 $150 - 250

905 Historic Costume Paper Doll Cut-Outs 1934 $45 - 65

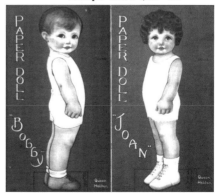

907 Paper Doll "Joan"/Paper Doll "Bobby" 1928 $50 - 75

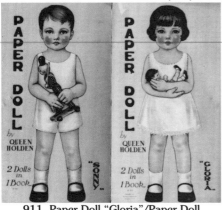

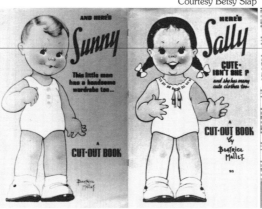

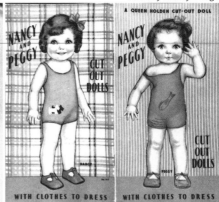

911 Paper Doll "Gloria"/Paper Doll "Sonny" 1930 $50 - 75

915 Here's Sally and Here's Sunny 1939 $60 - 75

917 Nancy and Peggy 1933 $35 - 50

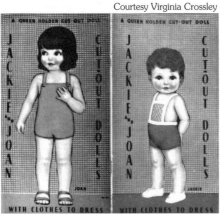

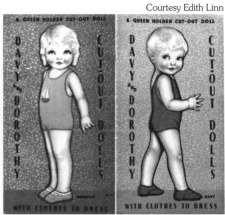

917 Jackie and Joan 1933 $35 - 50

917 Tommy and Jean 1933 $35 - 50

917 Davy and Dorothy 1933 $35 - 50

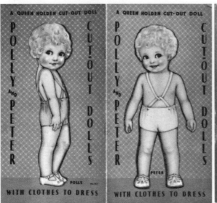

917 Polly and Peter 1933 $35 - 50

920 Baby Sister 1929 $65 - 80

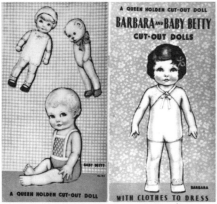

917 Barbara and Baby Betty 1933 $35 - 50

920 Baby Brother 1929 $65 - 80

931 Puppies and Kittens 1939 $75 - 160

932 They Stand Up 1936/39 $70 - 85 This is a reprint of **#983 Little Brothers and Little Sisters**. The dolls are redrawn.

920 The Twins 1932 $65 - 80 This large book is a reprint of **#920 Baby Sister** and **#920 Baby Brother**. The four dolls are redrawn. This book was also published as two separate books: **#920 The Twins** and **#920 Baby Brother and Baby Sister**.

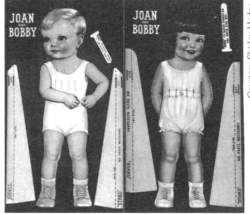

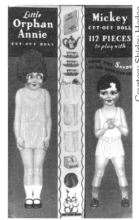

935 The New Joan and Bobby 1936
$85 - 100

938 Little Orphan Annie, Mickey 1931
$150 - 250

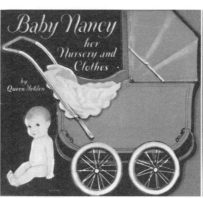

935 Walt Disney's Pinocchio 1939 $250 - 350

938 Baby Nancy 1931 $70 - 80

938 Little Orphan Annie 1934 $150 - 250

946 Raggedy Ann, Raggedy Andy 1935 $75 - 150

943 Baby Ann circa 1932 $75 - 100

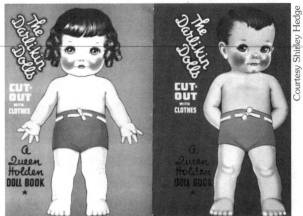

947 Winnie the Pooh, Christopher Robin 1935 $85 - 100

951 The Darlikin Dolls 1938 $90 - 150

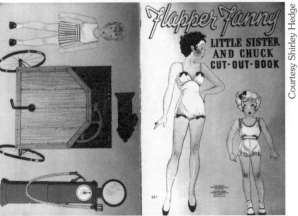

955 Peter Rabbit in 1939-40 catalog $65 - 90

957 Flapper Fanny 1938 $150 - 200

958 Life Size Paper Doll 1936 $100 - 125

960 Movie Starlets 1946 $125 - 150

961 Baby Dolls with Cloth-Like Clothes 1946 $40 - 50

961 Two Little Girls and They Grew and Grew 1949 $50 - 60

962 Betty Grable 1946 $175 - 275

962 Double Date 1949 $40 - 55

963 Blondie 1949 $100 - 150 This book has six pages of outfits. Reprint **#1191** has eight pages.

963 Margaret O'Brien 1946 $85 - 140

964 Lana Turner 1947 $175 - 275

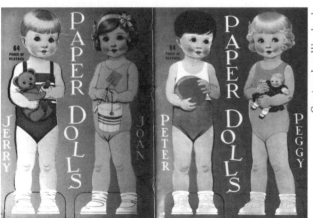

965 Paper Dolls 1935 $40 - 60

965 Ava Gardner 1949 $125 - 175 Six pages of outfits. Reprint **#1192** has eight pages.

966 Little Brother, Little Sister 1938 $75 - 120 This book has **#966** on one cover and **#968** on the reverse cover. The Whitman 1938 catalog has it numbered **#966**.

966 Portrait Girls 1947 $60 - 75

967 Blondie 1947 $100 - 150

967 Blondie 1948 $100 - 150

968 Elizabeth Taylor 1949 $125 - 200

968 Four Mothers and Their Babies 1941 $90 - 160 Right: Inside front cover

969 Baby Dolls 1945 $60 - 90

969 Walt Disney's Donald Duck and Clara Cluck
1937 $350 - 550

969 Dy-Dee Baby
Doll 1938 $90 - 150

969 Kewpies with Ragsy & Ritzy 1932 $200 - 350

970 Snow White and the Seven Dwarfs 1938 $175 - 250
Red background

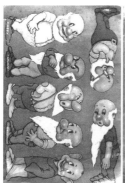

970 Walt Disney's Snow White and the Seven Dwarfs 1938 $150 - 225 Blue background

970 My Twin Babies 1940 $75 - 90 Right: Inside Covers

970 Margaret O'Brien 1944 $90 - 150

970 Dress Me! 1943 $75 - 100 Right: Inside front cover

Courtesy Emma Terry

970 June Allyson 1950 125 - 175

970 Four Baby Dolls 1953 $20 - 30

971 Sue the Cuddly Doll 1938 $60 - 80

971 Baby Dolls 1944 $40 - 50

971 Lola Talley
1942 $90 - 125

972 Betty Grable
1942 $150 - 175

972 Baby Sue 1944
$30 - 40

971 Little Brothers and Sisters
1953 $25 - 35

972 Dress Me! 1950
$40 - 55

Prices for the above two sets include the uncut
booklet tied to the doll's arm.

972 Look-Alike 1952 $35 - 50

973 Glamour Girl 1942 $50 - 70 Right: Inside cover

973 Career Girls 1944 $60 - 75

973 Elizabeth Taylor 1950 $125 - 200

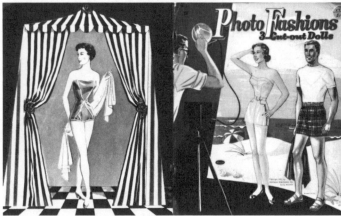

973 Photo Fashions 1953 $40 - 60

974 Girl Friends 1944 $50 - 75

Courtesy Shirley Hedge

974 Blondie 1950 $100 - 150

974 Hi Gals! 1953 $40 - 50

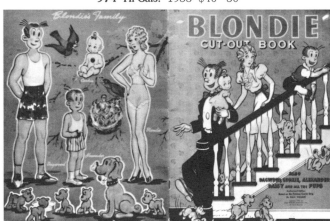

975 Mother and Daughter 1940 $50 - 65

975 Blondie 1943 $100 - 175 This **Blondie** book has been found with ten or eight inside pages.

Courtesy Emma Terry

975 Lana Turner 1945 $175 - 275

975 Three Sweet Baby Dolls 1954 $20 - 30

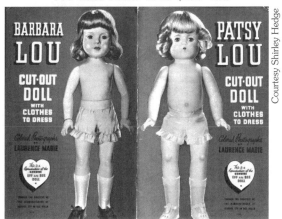

Courtesy Shirley Hedge

Courtesy Emma Terry

976 Patsy Lou, Barbara Lou 1939 $75 - 125

976 Bob Hope, Dorothy Lamour 1942 $200 - 300

976 Betty Grable 1943 $200 - 300

976 Playmates 1954 $25 - 35

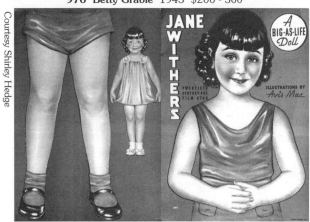

977 Jane Withers 1936 $150 - 200

977 Dolls That Walk 1939 $50 - 65

977 Dinah Shore 1943 $175 - 275

977 Sleeping Dolls 1945 $50 - 75

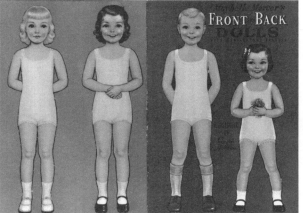

977 Mommy and Me 1954 $40 - 50

978 Maybelle Mercer's Front and Back Dolls 1939 $75 - 85

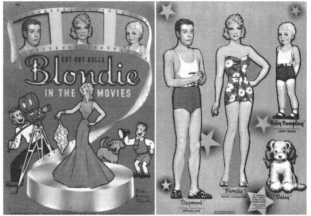

978 New! Sleeping Dolls $50 - 75 No date, listed as "new" in 1943 Whitman catalog.

979 Blondie in the Movies 1941 $125 - 200 Right: Inside front cover

979 Doll Town $100 - 160 No date, listed as "new" in 1938 catalog. Right: Inside pages

979 Surprise Party 1943 $60 - 75 Right: Inside front cover

979 Three Little Sisters 1943 $60 - 75

980 Baby Patsy 1934 $75 - 100 Right: Inside cover

980 Popeye 1937 $250 - 350

980 Judy Garland 1941 $150 - 200

981 Flossy Fair, Peter Fair 1933 $50 - 80

980 'Teen Gal Cut Out Dolls 1943 $50 - 75 Right: Inside front cover

981 Powers Models 1942 $100 - 175 Right: Inside front cover

981 Blondie 1944 $100 - 175

982 Blondie 1940 $100 - 175 Right: Inside front cover

982 Big and Little Cut-Out Dolls 1941/43 $50 - 75 Reprint of **#998 Big, Big Cut-Out Book** with one new doll, no wedding.

982 All Size Dolls 1945 $65 - 85

983 Cut-Out Dolls $40 - 55 No date, listed as "new" in 1939 catalog. Has new dolls drawn by Avis Mac. The original is **#965**.

983 Little Brothers and Little Sisters 1936 $80 - 100

983 3 Pretty Girls 1943 $60 - 75

984 Carolyn Lee 1942 $65 - 100 Right: Inside front cover

984 Judy and Jill 1945 $35 - 45 Some copies not numbered.

985 "All Star Movie Cut Outs" 1934 $150 - 250 Right: Inside pages

985 "All Star Movie Cut Outs" inside pages

985 Paper Doll Family 1937 $100 - 150 Right: Inside front cover

985 The Old Lady Who Lived in the Shoe 1940 $100 - 150
Right: Inside front cover

985 WACS and WAVES 1943 $100 - 175
Right: Inside front cover

985 Statuette Dolls 1946 $50 - 60   **986 Statuette Dolls** 1946 $45 - 50

986 Jane Withers 1941 $100 - 175

 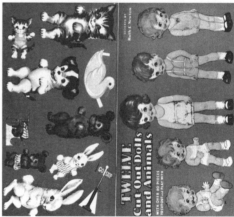

986 Ann Sheridan 1944 $175 - 275

987 Twelve Cut-Out Dolls and Animals 1934 $60 - 100

987 Bunny Boy 1938 $75 - 125

987 Daisy Bunny 1938 $75 - 125

  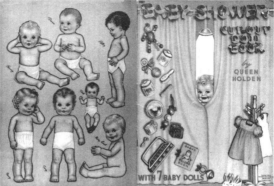 

987 Mother Goose 1937 $20 - 30   **987 Baby Shower** 1942 $75 - 100 Right: Inside front cover

987 Blondie 1945 $100 - 175   **988 Little Mother** 1940 $75 - 125 Right: Inside front cover

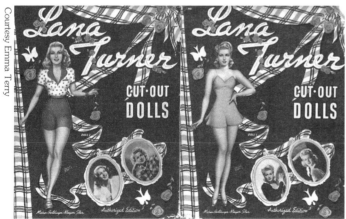 

988 Bobby Socks 1945 $60 - 75 Right: Inside front and back covers

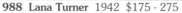  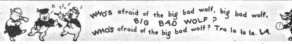

988 Lana Turner 1942 $175 - 275   **989 A Walt Disney Silly Symphony Cut-Out** 1933 $300 - 500

989 Jane Withers 1940 $100 - 175

989 Betty Grable 1941 $175 - 275

989 6 Cut Out Dolls 1943 $35 - 50 Right: Inside front cover

990 Rainbow Dolls 1934 $75 - 100

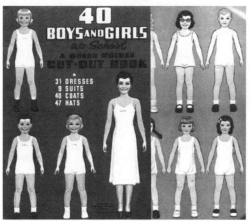

990 40 Boys and Girls at School 1939 $60 - 90

990 Joan's Wedding 1942 $75 - 100 Right: Inside front cover

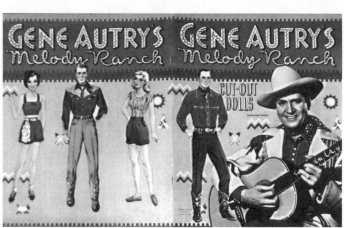

990 Three Little Girls 1945 $60 - 75

990 Gene Autry's Melody Ranch 1950 $100 - 150

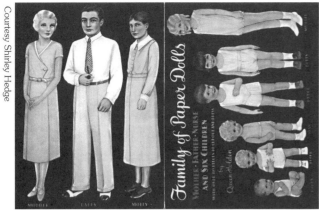

991 Family of Paper Dolls $75 - 90 No date, published in 1932.

991 2 Large Dolls, Sonny with Gloria 1938 $50 - 75 This book has new dolls. The original is **#911**.

991 The Fanny Brice Baby Snooks Cut-Out Doll Book 1940 $200 - 350

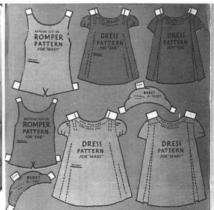

991 6 Movie Starlets 1942 $100 - 175 Right: Inside front cover

991 Hair-Do Dolls 1948 $50 - 75

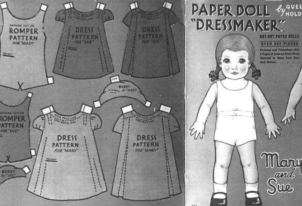

992 Patsy Ann 1939 $80 - 100 Right: Inside front cover

992 Paper Doll "Dressmaker" 1935 $50 - 65

992 Three Little Girls 1941 $75 - 90 Right: Inside front cover

992 Statuette Dolls 1943 $50 - 60

992 Gene Tierney 1947 $175 - 275

993 Alice in Wonderland 1933 $85 - 150

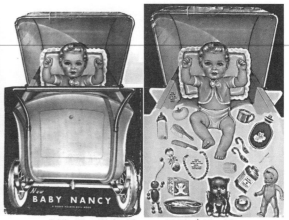

993 New Baby Nancy 1938 $65 - 90
Right: Inside front cover

993 Glamour Girl 1941 $50 - 75 Right: Inside front cover

993 Blondie 1945 $100 - 175

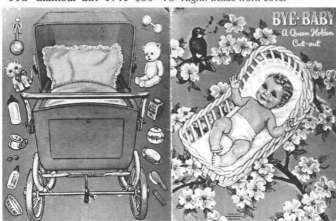

993 Bye-Baby 1950 $50 - 60

994 Little Women 1934 $85 - 100

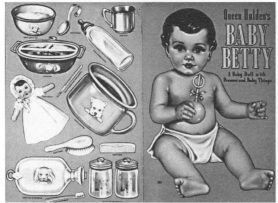

994 Baby Betty 1937 $85 - 100

994 Tot's Toggery 1942 $90 - 110

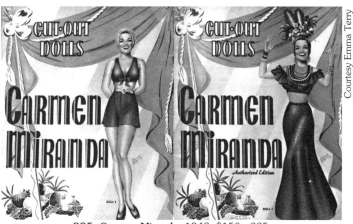

994 Betty and Bob 1948 $50 - 60

995 Carmen Miranda 1942 $150 - 225

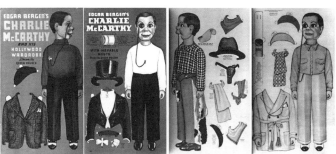

994 Frances Tipton Hunter's Paper Dolls 1943 $85 - 140
Right: Inside front cover

995 Edgar Bergen's Charlie McCarthy 1938 $125 - 250
Right: Inside pages

995 Fashion Cut-Out Dolls 1945 $50 - 70

996 Jane Withers 1938 $100 - 175

995 Playmates 1952 $30 - 40

995 Roy Rogers 1948 $125 - 175

49

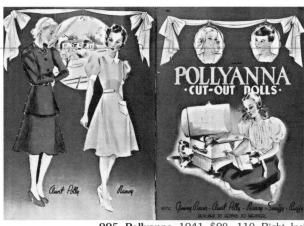

995 **Pollyanna** 1941 $90 - 110 Right: Inside front cover   **996** **Baby Sandy** 1940 $100 - 150

996 **Blondie** 1941 $100 - 200   **996** **"Sixteen"** 1943 $60 - 75 Reprint of **#980** 'Teen Gal. Dolls are redrawn.

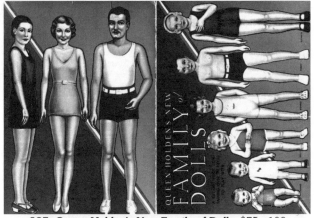

996 **Judy Garland** 1945 $200 - 300   **997** **Queen Holden's New Family of Dolls** $75 - 100
No date, listed as "new" in 1936 catalog.

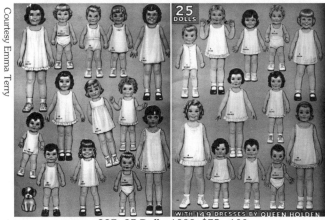
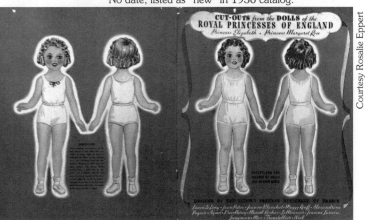

997 **25 Dolls** 1938 $75 - 100   **997** **Cut-Outs from the Dolls of the Royal Princesses Of England**
1940 $100 - 150

997 Tillie The Toiler 1942 $125 - 200

997 Carolyn Lee 1943 $75 - 100 Right: Inside front cover

998 Baby Jean 1938 $75 - 100 Right: inside front cover

998 Statuette Dolls 1942 $50 - 60

998 Five Paper Dolls 1935 $80 - 125

998 Big Big Cut Out Book 1941 $80 - 125 Right: inside covers

998 Picture Cut-Out Dolls 1946 $45 - 60

998 Roy Rogers and Dale Evans 1950 $125 - 175

998 Teen-Time $30 - 40 No date, pictured in the 1953 catalog.

999 Judy Garland 1940 $75 - 100 Right: Inside front cover

1002 This Is Margie 1939 $45 - 60   **1002** This Is Peggy 1939 $45 - 60   **1002** This Is Patsy 1939 $45 - 60

1002 This Is Dotty 1939
$45 - 60

1002 This Is Bunny 1939
$45 - 60

The five **#1002** books were also sold together in a box as **#2148**. The five dolls were also sold as heavy statuette dolls in separate box sets **#3967** or **#3037**. Each book has the same back cover as shown with **#1002 This Is Margie**. All five books are reprints of the five Dionne Quintuplet paper doll books **#1055** and contain the original clothes.

1011 My Baby Book 1942 $40 - 50

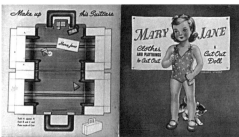

1010 Mary Jane 1939/41 $45 - 60

1010 Sally Ann 1939/41 $45 - 60
1010 Mary Lee 1939/41 $45 - 60

1010 Betty Sue 1939/41 $45 - 60
1010 Patty Lou 1939/41 $45 - 60

Each book has the same back cover shown with **#1010 Mary Jane**. All are reprints of the Dionne Quintuplet doll books **#1055** and contain the original clothes.

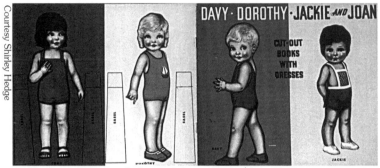

1013 Davy, Dorothy, Jackie and Joan 1933 $40 - 60 There are two other books in this series with **#1013** as their number. All three books are reprints of the **#917** books. However, these three books have the dolls and outfits printed in reverse.

1016 Betty Brewer 1942 $60 - 80

1016 Virginia Weidler 1942 $60 - 80

1016 Cora Sue Collins 1942 $60 - 80

1040 Play Dollies 1920's $65 - 90

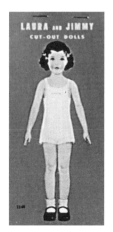

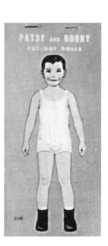

1146 Laura and Jimmy 1939 $16 - 23
1146 Patsy and Donny 1939 $16 - 23

These books are from a series of 12, all with the number **1146**. They are reprinted from **#990 40 Boys and Girls at School**.

1035 Roy Rogers' Double-R-Bar Ranch 1955 $75 - 100 Right: Inside pages   **1055 Annette** 1936 $80 - 120

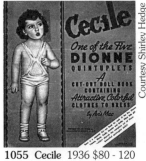

1055 Cecile 1936 $80 - 120   **1055 Emilie** 1936 $80 - 120   **1055 Marie** 1936 $80 - 120   **1055 Yvonne** 1936 $80 - 120

All **1055** Books have the same back cover.

1171 Jane Powell 1952 $125 - 175   **1171 Jane Powell** 1953 $125 - 175

1172 Baby Dolls 1950 $40 - 50   **1172 Roy Rogers and Dale Evans** 1952 $125 - 175

1173 The Cradle Crowd 1948 $50 - 75

1173 June Allyson 1953 $125 - 175

1174 Three Sisters 1942 $50 - 65

1174 Blondie 1953 $100 - 150

1175 Mary Hartline 1953 $100 - 175

1176 Nursery School Dolls 1953 $35 - 50

1177 Elizabeth Taylor 1953 $125 - 200

1178 Cloth-Like Clothes 1949 $60 - 75

1178 Debbie Reynolds 1953 $125 - 175

1179 Baby Cut-out Dolls 1949 $40 - 50 Right: Inside front cover

1179 Doris Day 1954 $125 - 175

1180 Sandra and Sue 1948 $45 - 50

1180 Blondie 1954 $100 - 150

1181 Sally 1950 $75 - 90

1181 We're A Family 1954 $40 - 50

1182 Brother and Sister 1950 $40 - 50

1184 Jimmy and Jane visit Gene Autry at Melody Ranch
1951 $90 - 125

56

1185 Jane Powell 1951 $125 - 175

1186 Roy Rogers and Dale Evans 1950 $125 - 175
A similar version came with a double cover

1187 Bridal Party 1950 $75 - 100

1188 Make Judy Laugh 1952 $35 - 50

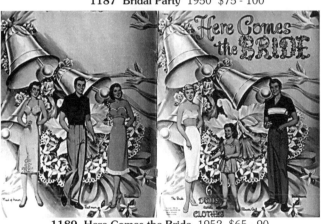

1189 Here Comes the Bride 1952 $65 - 90

1426 Wendy's Wardrobe Stencils 1965 $25 - 30

1502 Barbie 1990 $12 - 15 Right: Inside front cover

1502-1 Barbie 1990 $12 - 15 Right: Inside front cover

1500 Perfume Pretty Barbie 1988 $12 - 15
Right: Inside front and back covers

1501 Bride & Groom 1988 $12 - 15
Right: Inside front and back covers

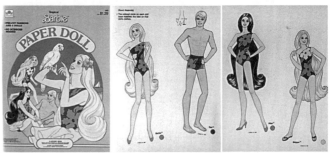

1502-3 Barbie 1992 $12 - 15
Right: Inside front cover

1503 Lil Miss 'n Me 1990 $8 - 10 Right: Inside front cover

1506 Quints 1990 $12 - 15 Right: Inside front cover

1523 Tropical Barbie 1986 $12 - 15 Right: Inside front and back covers

1523-2 Barbie 1990 $12 - 15
Right: Inside front cover

1524 Jem 1986 $10 - 12 Right: Inside front and back cover

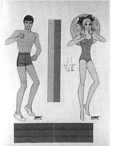

1528 Barbie and the Rockers 1986 $15 - 18 Right: Inside front and back cover

1532 Punky Brewster 1986 $12 - 15 Right: Inside front cover

1535 My Buddy 1986 $8 - 10 Right: Inside front and back cover   **1537 Jewel Secrets Barbie** 1987 $12 - 15 Right: Inside front and back cover

1537-2 Super Star Barbie 1989 $12 - 15 Right: Inside front cover   **1540 Lady Lovely Locks** 1987 $12 - 15 Right: Inside front and back cover

1541 Hot Looks 1988 $10 - 12 Right: Inside front cover   **1542 Moon Dreamers** 1987 $10 - 12
Right: Inside front and back cover

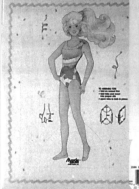

1544 Maxie 1989 $10 - 12 Right: Inside front cover   **1545 Walt Disney's Cinderella** 1989 $12 - 15 Right: Inside front cover

1674 Walt Disney's The Little Mermaid 1991 $12 - 15
Right: Inside front cover

1675 Disney's Beauty and the Beast 1991 $12 - 15
Right: Inside front cover

1682 Rock Stars 1992 $8 - 10

1688 Prom Night 1991 $8 - 10

1690 Barbie 1990 $10 - 12

1690-2 Barbie 1992 $10 - 12

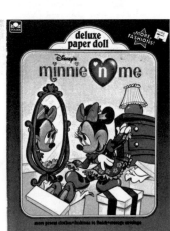

1691 Bride & Groom 1990 $12 - 15

1692 Disney's Minnie 'N Me 1990 $12 - 15

1693 Miss America (black) 1990 $10 - 12

1693 Miss America (white) 1990/91 $10 - 12

1695 Barbie 1991 $10 - 12

1695-1 Barbie 1991 $10 - 12

1697 Walt Disney's Snow White and the Seven Dwarfs 1991 $10 - 12

1698 Jim Henson's Muppet Babies, Baby Miss Piggy 1991 $8 - 10

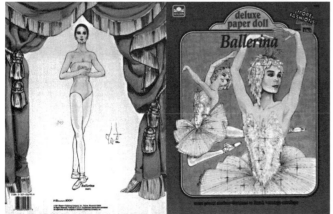

1699 Ballerina 1991 $8 - 10

1729 Christmas Bears 1984 $8 - 10 Right: Inside front cover

1731 Barbie Christmas Time 1984 $12 - 15 Right: Inside front cover   **1836 Barbie and Skipper** 1980 $12 - 18 Right: Inside front cover

1836-32 Starr 1981 $10 - 12 Right: Inside front and back covers   **1836-43 Pink & Pretty Barbie** 1983 $12 - 15
Right: Inside front cover

1837 The Ginghams 1980 $10 - 15 Right: Inside covers

1837-32 Bride and Groom 1981 $10 - 15 Right: Inside front covers   **1837-44 Best Friends** 1983 $10 - 12
Right: Inside front cover

1838 Raggedy Ann & Andy Circus Play Day 1980 $12 - 18
Right: Inside front cover

1838-32 The Four Bears 1981 $8 - 10 Right: Inside front cover

1838-44 The Original Monchhichi 1983 $8 - 10
Right: Inside front cover

 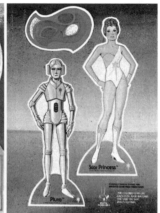

1839 Star Princess and Pluta 1979 $12 - 15 Right: Inside front cover

1839-32 Fashion Show 1981 $10 - 12 Right: Inside front cover

1839-41 Show Biz 1983 $8 - 10 Right: Inside front cover

1901 5 Dolls and Scissors 1954 $35 - 45

1941 Rub-a-Dub Dolly 1977 $8 - 10

1942 Canned Beans 1977 $8 - 10

1943 Daisy 1978 $8 - 10

1944 Wispy Walker 1976 $8 - 10

1947-2 Annabelle 1980 $8 - 10

Courtesy Virginia Crossley

1948 Here's The Bride 1960 $50 - 75 Right: Inside pages

1948 Janet Lennon 1961 $65 - 80   **1948 Debbie Reynolds** 1962 $100 - 125   **1948 National Velvet** 1962 $60 - 75

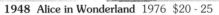

1948 Alice in Wonderland 1976 $20 - 25   **1949** Baby Tender Love   **1949** Denim Deb 1977 $8 - 10

1974 $10 - 12

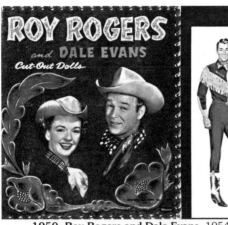

Courtesy Emma Terry

1950 Roy Rogers and Dale Evans 1954 $125 - 175   **1950** Roy Rogers and Dale Evans 1956 $125 - 175

1950 Roy Rogers, Dale Evans and Dusty 1957 $125 - 175   **1950** Baby Beans and Pets 1978 $8 - 10

1950 Walt Disney's Snow White 1974 $20 - 25   **1951** Little Ballerina 1959 $40 - 50

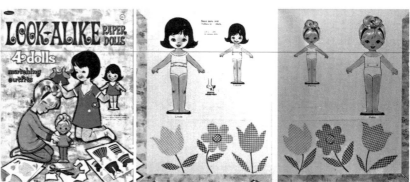

1951 Elizabeth Taylor 1955 $125 - 200   **1951** Look-Alike 1967 $25 - 30 Right: Inside back and front cover

1951 Barbie Goin' Camping 1974 $15 - 20   **1952** Doris Day 1955 $100 - 150

1952 Doris Day 1956 $100 - 150   **1952** Heidi 1966 $25 - 30 Right: Inside front cover

1952 Shrinkin' Violette 1965 $20 - 25 Right: Inside front cover   **1952** Tina and Trudy 1967 $40 - 50

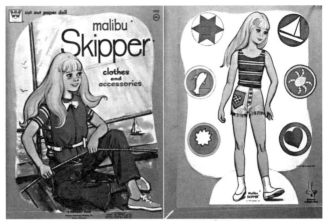

1952 **Teen Model** 1968 $18 - 25 Right: Inside back and front cover

1952 **Malibu Skipper** 1973 $15 - 25
Issued as a folder, then as a book. Right: Inside

1953 **Bridal Party** 1955 $75 - 125 Right: Inside

1953 **Here's the Bride** 1956 $65 - 90

1953 **The Bride** 1957 $65 - 90 Right: Inside pages

1953 **Annette** 1964 $75 - 100 Right: Inside

1953 **Wedding Dolls** 1958 $60 - 85 Right: Inside pages

1953 **Tammy and Pepper** 1966 $50 - 65

1953 Mod Matchmates 1970 $8 - 12   **1953 Cry Baby Beans** 1973 $8 - 10 Right: Inside

1954 Oklahoma! 1956 $100 - 175   **1954 Wishnik Cut-Outs** 1966 $25 - 35

1954 Petticoat Junction 1964 $85 - 100   **1954 Secret Sue** 1967 $20 - 25

1954 Playhouse Kiddles 1971 $25 - 35   **1954 Barbie's Boutique** 1973 $15 - 25 Right: Inside

1955 Debbie Reynolds 1955 $100 - 150   **1955** Debbie Reynolds 1957 $100 - 150

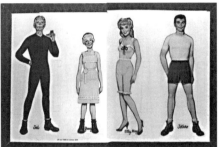 

1955 Tiny Tots 1958 $40 - 50 Right: Inside   **1955** The Beverly Hillbillies 1964 $85 - 100 Right: Inside

 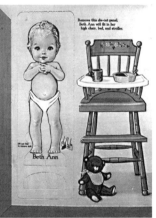

1955 Haley Mills in "That Darn Cat" 1965 $65 - 90   **1955** Beth Ann 1970 $30 - 40

1955 Malibu Francie 1973 $25 - 35 First issued folder style, then as a book.   **1956** June Allyson 1955 $100 - 150

1956 Janet Lennon 1959 $65 - 80   **1956** Debbie Reynolds 1960 $100 - 125   **1956** Baby Brother and Sister 1961 $25 - 35

1956 Annette 1962 $75 - 100   **1956** Connie Francis 1963 $65 - 80   **1956** Sandy and Sue 1963 $20 - 30

  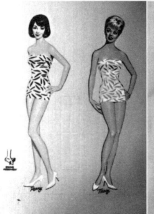 

1956 Nurses Three 1965 $40 - 50 Right: Inside

1956 Storybook Kiddles Sweethearts 1969 $35 - 45 Right: Inside

1956 Rock Flowers 1972 $8 - 10 Right: Inside pages   1956 Yellowstone Kelley 1975 $8 - 10

1957 4 Ballet Dancing Dolls 1955 $25 - 35   1957 Ballet Dolls 1957 $30 - 35

1957 Walt Disney's Mouseketeer Linda 1958 $50 - 75   1957 Barbie and Skipper 1964 $35 - 50

1957 Baby Bumpkins 1969 $12 - 18   1957 Tearful Baby Tender Love 1974 $12 - 18

1958 Walt Disney's Mouseketeer Annette
1956 $80 - 100

1958 Trixie Belden 1958 $60 - 80

1958 National Velvet 1961 $60 - 75

1958 Bride Cut-Outs 1963 $50 - 75

1958 Bridal Party 1964 $50 - 75

1958 Bridal Cut-Outs 1965 $50 - 75

1958 Happy Bride 1967 $50 - 75

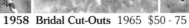

1958 Bride and Groom 1968 $50 - 75 Right: Inside

1958 Dollikin 1971 $10 - 12 Right: Inside

1958 Lazy Dazy 1973 $8 - 10 Right: Inside

1958 Itsy Bitsy Beans 1975 $8 - 10

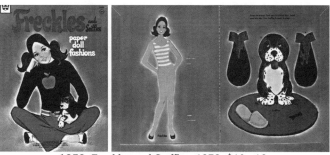

1959 The Munsters 1966 $100 - 150 Right: Inside

1959 Freckles and Sniffles 1972 $10 - 12

First issued folder style, then as a book.

1959 Walt Disney's Mouseketeer Cut-Outs 1957 $65 - 90

Right: Inside

1959 Charmin' Chatty 1964 $35 - 50

1959 Bundle of Love 1969 $12 - 18

1960 Annie Oakley 1956 $75 - 90

1960 Hayley Mills "The Moon-Spinners" 1964 $65 - 90 Right: Inside

1960 Baby Secret 1966 $20 - 25 Right: Inside

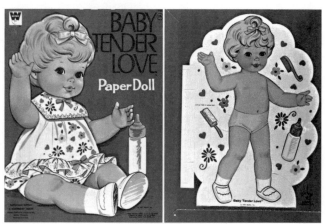

1960 Dolly Dears 1969 $10 - 15 Right: Inside

1960 Baby Tender Love 1971 $12 - 15

1961 Baby Brother and Sister 1958 $25 - 35

1961 Chatty Cathy 1963 $35 - 50
"© 1960/1962" also on cover

1961 Chatty Cathy 1964 $35 - 50

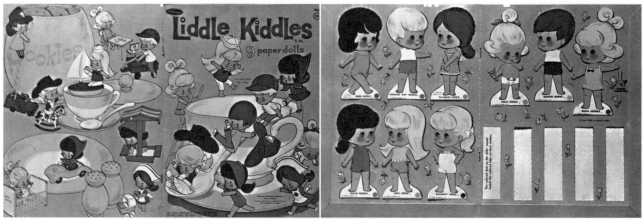

1961 Liddle Kiddles 1966 $50 - 65 Right: Inside

1961 Little Red Riding Hood 1972 $10 - 12 **1961** Jean Jeans 1975 $8 - 12

1962 Natalie Wood 1957 $100 - 175 **1962** Barbie 1963 $60 - 80

1962 Midge 1963 $60 - 80 Right: Inside **1962** Storybook Paper Dolls 1965 $15 - 20 Right: Inside

1962 Ballet Cut-Outs 1964 $15 - 25 Right: Inside

1962 Now Playing Ballet Paper Dolls 1966 $15 - 25

1962 Ballet Dancers 1968 $15 - 25

1962 Goldilocks 1972 $10 - 12

1962 Raggedy Ann and Andy 1974 $15 - 18

1963 Dinah Shore 1958 $100 - 125

1963 Barbie Doll 1962 $60 - 90 Right: Inside

1963 Little Ballerina 1961 $18 - 25 Right: Inside

1963 Lucy 1964 $80 - 100

1963 Dolly Darlings 1966 $12 - 15

1963 Sweet April 1973 $10 - 12   **1964** Janet Lennon 1958 $65 - 80   **1964** Carol Heiss 1961 $65 - 75

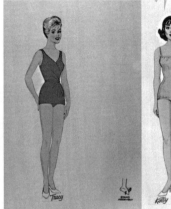

1964 Dorothy Provine 1962 $60 - 70   **1964** Nurses Three 1964 $40 - 50 Right: Inside

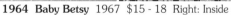
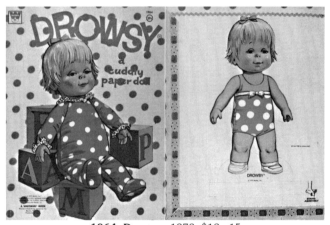

1964 Baby Betsy 1967 $15 - 18 Right: Inside   **1964** Drowsey 1973 $12 - 15
First issued folder style, then as a book. Right: Inside

1964 6 Paper Doll Playmates 1968 $10 - 15 Right: Inside   **1964** Mary Dress 'N' Play 1969 $12 - 15

1964 Littlest First Born 1971 $10 - 12 Right: Inside 1965 Charmin' Chatty Play Hats
1964 $12 - 15

1965 Dodi 1966 $30 - 40

1965 Wishnik Cut-Outs 1965 $30 - 40   1965 Pee Wee Cut-Outs 1967 $30 - 40 Right: Inside

1966 Hayley Mills in "Summer Magic" 1963 $65 - 90

1965 Tiny Tot Shop 1969 $40 - 50

1965 Valerie 1971 $12 - 15

1966 Marge and Gower Champion 1959 $100 - 150

1966 Pretty Belles 1965 $15 - 18   **1966** Peachy & Her Puppets 1974 $8 - 10

1967 Newborn Thumbelina 1969 $10 - 12   **1967** Bridal Paper Dolls 1971 $12 - 15

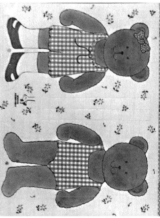

1967 Walt Disney's Mary Poppins
1973 $35 - 50   **1967** Teddy Bears 1978 $10 - 12   **1968** Sweet Swingers 1969 $8 - 10

1968 Pat Boone 1959 $60 - 80   **1968** Little Dancers 1970 $8 - 10   **1968** Matchin' Mods 1973 $8 - 10

1969 Tiny Tots 1959 $25 - 40

1969 Julie's Wedding 1961 $50 - 75
Same dolls as **#4723 Wedding Day Dolls** with new clothes

1969 Annette in Hawaii
1961 $75 - 100

1969 Tiny Toddles
1969 $15 - 18

1969 Baby Kim Cut-Outs
1962 $15 - 20

1969 Betsy McCall
1971 $25 - 35

Color Plates Identification

Description of Lowe Color Pages 81 and 84

As stated in the Introduction, reprints containing the same dolls as the originals are not pictured but are included in the lists. Whitman reprints with redrawn dolls are covered in the text. A few of the Lowe reprints with redrawn dolls are also covered in the text, but space limitations didn't allow all of these books to be shown. Because the author believes they should be shown, they are grouped together on two of the color pages. They are listed here with their stock numbers so they can be found in the Lowe list, which in turn will show which original book they are derived from.

Lowe Color Page 81 (left to right)
Top Row: #128 The Twins Bob and Jean, #1027 The Turnabout Twins, #1259 Fritzi Ritz, #1022 Clothes Make A Lady, #1241 Clothes Make A Lady
2nd Row: #148 The Carol and Jerry Cut Out Doll Book, #1024 Bumpity Bess, #1327 Ann and Betty, #1351 The First Seven Years of Penny, #522 Sonny & Sue

3rd Row: #1021 Mary Ann Grows Up, #124 Nancy - Judy, #1044 College Girls, #1327 Baby Buggy, #989 Nora Drake
4th Row: #2705 Kit and Kapoodle, #1280 Diaper Doll, #524 Betsy and Bill (opened out), #528 Betsy and Bill, #3942 My Dolly Kisses, #133 My Big Dolls
Bottom Row: #523-1 Four Jolly Friends, #525-1 Three Little Sisters, #525-2 Stand-Out, #1443 Betty Bo Peep, #1047 Maid of Board, #2730 Polly and Her Dolly

The values of reprints with redrawn dolls are lower than the original books except in the following cases: Fritzi Ritz $80 - 100, Sweetie Pie $50 - 65, Trixie $50 - 65, and Nora Drake $65 - 90 (in this book the original non-celebrity dolls were used as Nora Drake).

Lowe Color Page 84 (left to right)
Top Row: #1857 Vacation, #2744 Baby Sister, #2783 Dolltime (opened out), #1517 Fashion Models (opened out), #1014 10" Doll.
2nd Row: #4220C Baby Doll, #2482 Sweetie Pie, #2761 Baby Doll (opened out), #5909J Lace-On Clothes Dolly, #2777 Betty
3rd Row: #2769 Dollies Go To School (opened out), #2787 Little Miss Muffet, #2754 Betty Buttercup, #2778 Dress-Up Dolls (opened out)
4th Row: #4502 I'm Mimi/I'm Judy (opened out), #6901 Fairy Princess, #3920 Trixie, #4503 I'm Ginny, #4501 I'm Amy
Bottom Row: #1378 Teen Queens, #1367 TV Star Time, #1255 Fairy Princess, #2796 Betty Buttercup, #2528 Annie the Action Doll

CAREER GIRLS Paper Dolls

Nurse
Airline Hostess
Model
Schoolteacher
Business girl
Movie Star

The **8 AGES of JUDY** Eight Cut-Out Paper Dolls

The **Baby Show** 25 Dolls

TURNABOUT DOLLS A 22 DOLL CUT-OUT BOOK

PiXiE DOLL and PUP to push out and dress

30 INCH STAND-UP DOLL **Kathy** PUSH-OUT CARDBOARD CLOTHES

MERCURY RECORD and TV STAR **Patti Page** 2 Statuette Dolls & Fabulous Wardrobe

A Box of **10 CUT-OUT DOLLS** With Dresses

10 CUT-OUT DOLLS

10 CUT-OUT DOLLS

THE **Bobbsey Twins** PAPER DOLLS

Freddie Flossie

Dude Ranch TURNABOUT PAPER DOLLS

PRUDENCE

Helene

Anne

Mary

Carol

Nell

Jonan

JANE

SUSAN

Louise

Little **BEAR** to DRESS

Little **DOG** to DRESS

Little **PIG** to DRESS

Little **KITTEN** to DRESS

The **BOB CUMMINGS FASHION MODELS** 4 Statuette Dolls and Clothes

Baby Doll 10 INCH STATUETTE DOLL and CLOTHES

GIRLS IN UNIFORM PAPER DOLLS

Betty Bo-Peep cut out doll By mardy

Gloria's **MAKE-UP** PAPER DOLL BOOK LIFE-SIZE HEAD AND HANDS TO MAKE-UP ON BACK COVER — AND DRESSES. 10 CUT OUT INSIDE

Judy and **Jack** CUT-OUT DOLLS 40 PAGES OF DOLLS AND DRESSES

DOWN on the FARM
Dolls with clothes

Clothes Crazy
TWO POSES FOR EACH DOLL

DICK the SAILOR

HARRY the SOLDIER

TOM the AVIATOR

Rockabye Babies PAPER DOLLS

Rock-a-Bye Babies

Rock-a-Bye Babies on the tree top
Hold the book up and the cradle will rock
Open the page and for each little doll
Down will come baby clothes, cradle and all

Toni
HAIR-DO Cut-Out Dolls

LOOK FOR the Toni Dolls ON BACK COVER

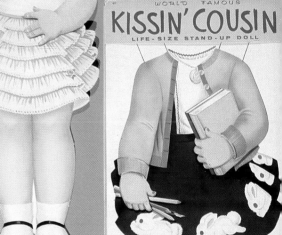

TOM AND HIS TOYS

BABY

FARMER FRED AND HIS FRIENDS

JANIE AND HER DOLL

MARY AND HER TOYS

GOLDILOCKS
STATUETTE AND PAPER DOLLS

Kay and Kim
With Clothes To Trim
ACTIVITY DOLL BOOK
with colored gummed pages
DOLLS ON BACK COVER

Little Cousins
CUT-OUT DOLLS
with 32 PAGES OF DRESSES
Ginny

The Five Little Peppers
PAPER DOLLS

Baby Anne
LOTS OF PRETTY CLOTHES

KISSIN' COUSIN
WORLD FAMOUS
LIFE-SIZE STAND-UP DOLL

Bride and Groom Dolls

JUNIOR MISSES
TEEN-AGE DOLLS
CLOTHES FOR SCHOOL   PLAY · PROMS and DATES

POLLY PAL
PAPER DOLL BOOK

TWINKLE TWINS 4 YEARS OLD
TWINKLE TWINS 10 YEARS OLD

PLAYTIME PALS
PAPER DOLL   Cut-Out Book
Jack   Susan   Ann

PAT
THE "STAND-UP" DOLL
With FRONT and BACK DRESSES

Three Little Maids
from school are we
3 STATUETTE DOLLS AND THEIR CLOTHES

Let's Play House
Furniture and Dolls with Dresses

LITTLE BROTHER LITTLE SISTER

6 DOLLS BY QUEEN HOLDEN

CUTOUT DOLL & DRESSES

PATSY BOBBY JEAN

BABY GROWS UP Cut-Out Dolls

TRUDY PHILLIPS and Her Crowd

4 CUT-OUT DOLLS

THE BIG BOOK

10 Dolls and 100 Dresses

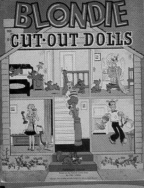

BLONDIE CUT-OUT DOLLS

BETTY and BILLY

Two Cut-Out Dolls

Frances Tipton Hunter's

PAPER DOLLS

Billy Patsy Susie Jerry Bob Judy

Big Big Cut Out Book

DRESS ME!

CUT-OUT DOLLS

By QUEEN HOLDEN

Girl friends

Bobby Socks

CUT-OUT DOLLS

DOLLS THAT WALK

THEY WALK • THEY DANCE • THEY PLAY

WALK-A-BIT Dolls

Joan's Wedding

MAGIC STAY-ON CLOTHES!

for Peter and Polly

Clothes Stay on Without Tabs

NO TABS

TWO DOLLS

The **OLD LADY** Who Lived **IN THE SHOE** Cutouts

24 PAPER DOLLS WITH CLOTHES

Surprise Party CUT-OUT DOLLS

Our **WAC** Joan

A STAND-UP DOLL COMPLETE WITH DRESSES

SOLDIER • WAC • NURSE WAVE • SAILOR

Our **WAVE** Patsy

A STAND-UP DOLL COMPLETE WITH DRESSES

Our **SAILOR** Bob

A STAND-UP DOLL COMPLETE WITH DRESSES

SOLDIER • WAC • NURSE WAVE • SAILOR

Our **SOLDIER** Jim

A STAND-UP DOLL COMPLETE WITH DRESSES

SOLDIER • WAC • NURSE WAVE • SAILOR

DOLL TOWN

A DOUBLE VALUE CUT-OUT-BOOK

40 PAGES OF DOLLS AND DRESSES All in Color

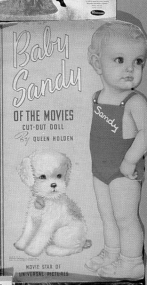

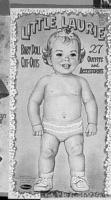

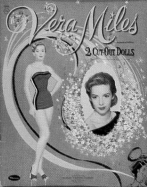

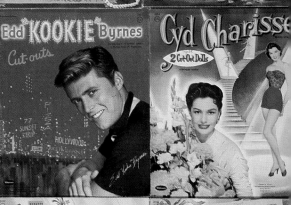

1969 Hi! I'm Skipper
1973 $20 - 30

1970 Dinah Shore and George Montgomery
1959 $100 - 150

1970 Baby Bonnie 1960 $20 - 30

1970 My Raggedy Ann Doll Book
1967 $20 - 30

1970 Raggedy Ann and Andy 1969 $20 - 30

1970 Little Lulu 1971 $35 - 50

1970 Lydia 1977 $12 - 15

1971 Annette 1960 $75 - 100

1971 Barbie and Ken 1962 $60 - 90

1971 My Susie Doll Book
1968 $15 - 18

1971 Pretty Pat 1969 $15 - 18

1971 Nancy 1971 $30 - 45

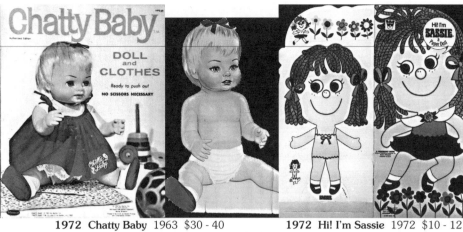

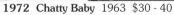

1972 Chatty Baby 1963 $30 - 40   **1972** Hi! I'm Sassie 1972 $10 - 12   **1973** Tini Tot Mods 1969 $10 - 12

1972 Peggy and Me 1968 $10 - 12

1973 Mrs. Beasley
1970 $20 - 30

1973 Baby Beans
1973 $10 - 12

1973 Tender Love 'n Kisses 1978 $10 - 12   **1974** Walt Disney's Mouseketeers 1963 $65 - 90

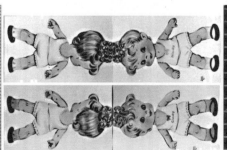

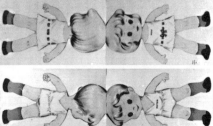

1974 Cindy and Mindy 1960 $25 - 35   **1974** Peter and Pam 1961 $25 - 35

1974 Cuddly Baby 1969 $12 - 15
Like **#1975** w/white baby

1974 Groovy P.J.
1972 $15 - 20

1974 Hi! I'm Valerie
1974 $12 - 15

1975 Baby Grows Up 1955 $50 - 60

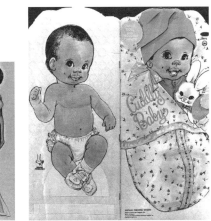

1975 The Nurses 1963 $40 - 60

1975 Cuddly Baby 1969 $15 - 18
Like **#1974**, w/ black baby

1975 Pos 'n' Barbie 1972 $18 - 30

1975 Crissy 1973 $15 - 20

1976 It's A Date 1956 $40 - 60

1976 Barbie, Ken and Midge 1963 $50 - 75

1976 Barbie Costume Dolls 1964 $75 - 90

 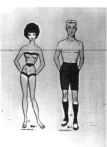    

1976 Barbie and Ken 1963 $50 - 75
Dolls are dated 1962 and are the same as dolls
in **#4797** dated 1962, but clothes are different.

1976 Barbie, Christie, Stacey 1968 $40 - 50

1976 Barbie, Skipper, Scooter 1966 $45 - 60

1976 Barbie Has a New Look! 1967 $45 - 55

1976 Barbie 1969 $40 - 50

1976 Barbie and Ken 1970 $35 - 40

1976 Groovy World of Barbie 1971 $35 - 40

1976 The Brady Bunch 1973 $40 - 50

1976 The Sunshine Family 1974 $15 - 18

1976 Honey Hill Bunch 1977 $10 - 12

1977 Doris Day Doll
1957 $100 - 150

1976-2 Strawberry Sue 1979 $10 - 20

1976-3 Miss America 1979 $12 - 15

1977 Raggedy Ann 1970 $15 - 20

1977 Tiny Tots 1967 $20 - 25

1977 Tini Go-Along 1969 $12 - 15

1977 Raggedy Ann and Andy 1972 $15 - 18
Dolls are same as **#1979**, clothes are different.

1977 Pippi Longstocking 1974 $25 - 30

1977-23 Raggedy Ann and Andy 1980 $12 - 15

1977-24 Walt Disney's Winnie-The-Pooh and Friends 1980 $25 - 30

1978 Baby Show 1957 $45 - 60

1978 Lovable Babies 1966 $18 - 25

1978 "Baby's Hungry" 1968 $18 - 20

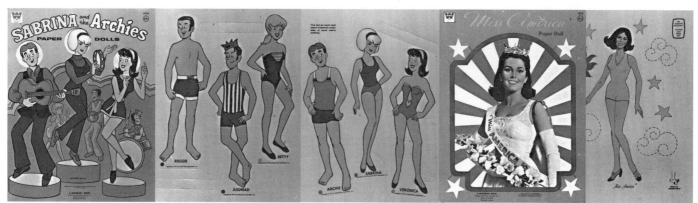

1978 Sabrina and the Archies 1971 $35 - 50

1978 Miss America 1973 $15 - 18

1978 The Happy Family 1977 $12 - 15

1978-3 My Best Friend 1980 $8 - 10

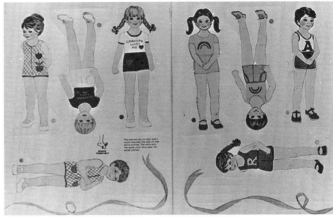

1978-24 Neighborhood Kids 1980 $8 - 10

1978-24 Neighborhood Kids

1979 Lennon Sisters 1957 $75 - 100   **1979** Lennon Sisters 1958 $75 - 100

1979 Raggedy Ann and Andy 1966 $25 - 30

1979 Green Acres 1967 $60 - 85   **1979** Nancy and Sluggo 1974 $15 - 20

95

1979 Little Lulu 1973 $25 - 35

1979 Walt Disney's Mickey & Minnie Steppin' Out 1977 $15 - 25

1979-2 Star Princess and Pluta 1979 $12 - 15

1980 Magic Stay-On Clothes for Peter and Polly 1956 $40 - 55

1980 Debbie Reynolds Doll 1958 $100 - 150

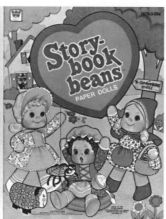

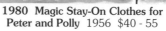

1979-3 Story-Book Beans 1980 $8 - 12 Right: Inside

1980 Meet Francie 1966 $30 - 40

1980 Tubsy 1968 $15 - 18

1980 Petal People 1969 $12 - 15   **1980** The Sunshine Family 1977 $12 - 15

1980 Dawn 1971 $15 - 20   **1980-2** Baby This 'n That 1979 $10 - 12   **1980-3** Super Teen Skipper 1980 $15 - 18

1981 Walt Disney's Sleeping Beauty 1959 $60 - 80   **1981** Walt Disney's It's a Small World 1966 $45 - 60

1981 Liddle Kiddles 1967 $50 - 65   **1981** Storybook Kiddles 1968 $50 - 60

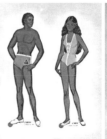
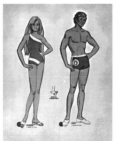

1981 P.J. Cover Girl 1971 $12 - 15   **1981** Barbie and Her Friends 1975 $20 - 25

1981 New 'N' Groovy P.J. 1970 $15 - 18

1981 Barbie's Sweet 16 1974 $20 - 25

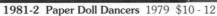
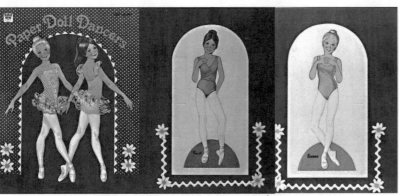

1981-2 Paper Doll Dancers 1979 $10 - 12

1981-3 Simply Sam 1980 $8 - 10

1982 The Flintstones 1961 $75 - 90
Large stand-ups of Fred, Barney, Wilma and Betty with outfits.
Right: Inside pages

1982 Walt Disney's Mary Poppins 1964 $40 - 50

1982 Walt Disney Presents Mary Poppins 1966 $40 - 50

1982 Chitty Chitty Bang Bang 1968 $40 - 50

1982 Babykins 1970 $10 - 12

1982 Josie and the Pussycats 1971 $30 - 40

1982 Francie 1973 $18 - 25
First issued folder style, then as a book.

1982 Baby Dreams 1976 $8 - 10

1982-1 Rosebud 1978 $8 - 10

1982-23 Teddy Bear Family 1980 $8 - 10

1982-31 Starr 1980 $10 - 15

1982-34 Pretty Changes Barbie 1981 $10 - 15

 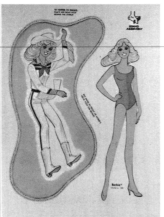 

1982-43 Western Barbie 1982 $10 - 15   **1982-44 Sunsational Malibu Barbie** 1983 $10 - 15

1982-45 Angel Face Barbie 1983 $10 - 15   **1982-46 Twirly Curls Barbie** 1983 $10 - 15

1982-47 Barbie "Fantasy" 1984 $10 - 15   **1982-48 Day-to-Night Barbie** 1985 $10 - 15

    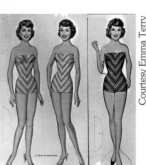

1982-49 Great Shape Barbie 1985 $10 - 15   **1983 The McGuire Sisters** 1959 $100 - 150

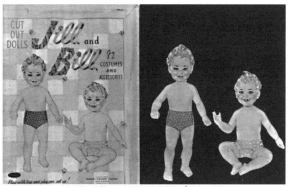

1983 Jill and Bill 1960 $20 - 25   **1983** The Lennon Sisters 1961 $75 - 100

1983 The Jetsons 1963 $60 - 75   **1983** Strawberry Sue 1973 $10 - 12

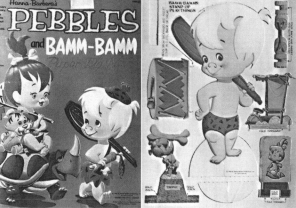

1983 Pebbles and Bamm-Bamm 1964 $40 - 60

1983 Pebbles and Bamm-Bamm 1965 $40 - 60   **1983** Pebbles and Bamm-Bamm 1966 $40 - 60

101

1983 Paper Dolls of Early America 1975 $8 - 10

1983 SuperStar Barbie 1977 $12 - 15

1983-34 Freckles & Sniffles 1981 $8 - 10

1983-35 We're Twins 1981 $8 - 10

1983-42 The Original Monchhichi 1982 $8 - 10

1983-43 Golden Dream Barbie 1982 $10 - 15

1983-44 Pink & Pretty Barbie 1983 $10 - 15

1983-45 Poochie 1983 $8 - 10

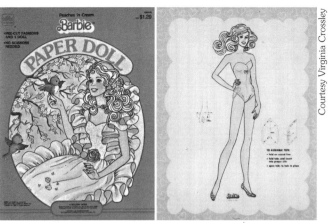

1983-46 Crystal Barbie 1984 $10 - 15   **1983-48 Peaches 'n Cream Barbie** 1985 $10 - 15

1983-50 The Heart Family 1985 $8 - 10   **1984 The Happy Family** 1960 $35 - 50

1984 Skipper 1965 $50 - 75   **1984 Peepul Pals** 1967 $25 - 30

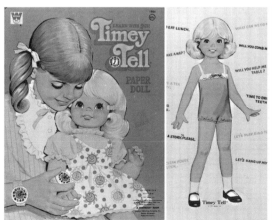

1984 Timey Tell 1971 $10 - 15   **1984 Quick Curl Barbie** 1973 $35 - 40

1984-44 Butterscotch 1983 $8 - 10

1984-46 Rainbow 1983 $8 - 10

1984-48 Wedding 1984 $8 - 10

1984-54 Princess of Power 1985 $8 - 10

1984-51 Rainbow Brite 1984 $8 - 10

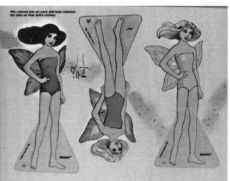

1984-53 Star Fairies 1985 $8 - 10

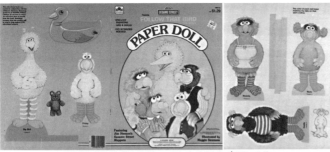

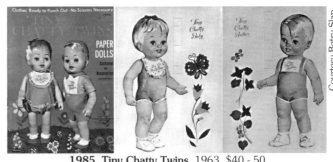

1984-55 Follow That Bird 1985 $12 - 15   **1985 Tiny Chatty Twins** 1963 $40 - 50

1985 My Doll Family 1955 $50 - 65   **1985 My Doll House Family** 1957 $50 - 65

1985 Charmin' Chatty 1964 $40 - 50   **1985 Skooter** 1965 $50 - 75

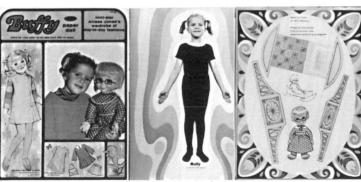

1985 Buffy 1968 $30 - 40   **1985 Buffy** 1969 $30 - 40

1985 Star Charmers 1970 $8 - 10   **1985 The Ginghams** 1976 $15 - 18

1985-34 Peek-A-Boo Baby 1981 $8 - 10   **1985-45 Bugs Bunny & Honey Bunny** 1983 $12 - 15

1985-46 My Very First Paper Doll 1983 $8 - 10   **1985-47 My Little Sister** 1983 $8 - 10

1985-48 Sesame Street Paper Doll Seasons 1984 $12 - 15   **1985-50 Princess Diana** 1985 $20 - 25

1985-49 Walt Disney's Mickey & Minnie 1983 $10 - 15   **1985-51 Barbie & Ken** 1984 $10 - 15

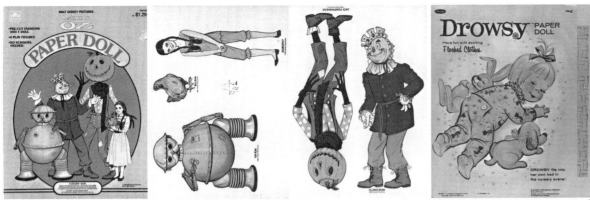

1985-60 Return To Oz 1985 $10 - 15   **1986 Drowsy** 1965 $15 - 20

1986 Francie and Casey 1967 $30 - 45   **1986 Playtime Pals** 1970 $10 - 12

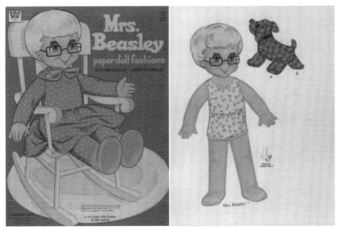
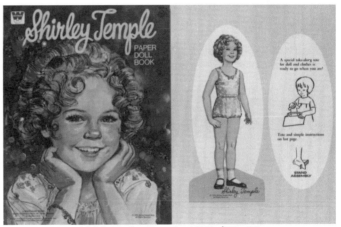

1986 Mrs. Beasley 1972 $20 - 25   **1986 Shirley Temple** 1976 $20 - 25

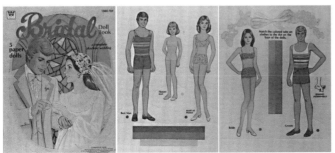

1986 Bridal Doll Book 1978 $15 - 18   **1987 Snow White and the Seven Dwarfs** not dated, but 1967 $50 - 65

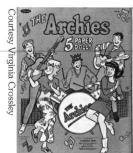
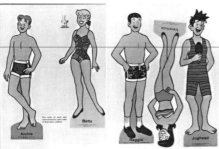

1987 The Archies 1969 $40 - 60

1987 World of Barbie 1971 $35 - 40

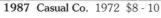

1987 Casual Co. 1972 $8 - 10

1987 Wizard of Oz 1976 $20 - 25

1987-31 Trixie Belden with Honey Wheeler 1981 $12 - 15

1987-33 The Ginghams Visit Grandma 1981 $12 - 18

1988 The Wedding Playbook 1960 $75 - 100

1988 Baby Go-Along 1968 $25 - 35

1987 Little Lulu and Tubby 1974 $15 - 25

1988 Tini Mods 1968 $8 - 12

1988 Winnie's Wardrobe 1966 $10 - 15

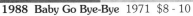
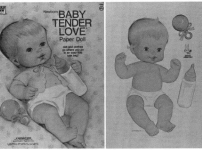

1988 Baby Go Bye-Bye 1971 $8 - 10   **1988** Newborn Baby Tender Love   **1988** Big Jim and Big Jack 1976 $10 - 15

1973 $12 - 15

1988-1 Raggedy Ann and Andy 1978 $12 - 15   **1989** Santa's Workshop 1960 $50 - 60

1989 Santa's Toyland Playbook 1962 $50 - 60   **1989** Bride and Groom 1970 $18 - 25

1989 Wedding Paper Dolls 1966 $30 - 40

109

1989 Bride Doll Book 1972 $18 - 25   **1989 Bridal Fashions** 1973 $18 - 25

1989 Barbie Fashion Originals 1976 $20 - 25

1990 The Nutcracker Ballet Cutouts 1960 $60 - 75

1990 Santa's Playbook 1964 $50 - 60

1990 Missy Go-Along 1970 $8 - 10

1990 Barbie Country Camper 1973 $20 - 25

1990 Growing Up Skipper 1976 $18 - 25

1990-1 Walt Disney's Daisy and Donald 1978 $15 - 25

1991 Lennon Sisters 1959 $75 - 100 Folder color is green or pink.

1991 Dennis the Menace Back-Yard Picnic 1960 $45 - 60

1991 Lucy and Her TV Family 1963 $85 - 125

1991 Patty Duke 1964 $40 - 50

1991 Patty Duke 1965 $40 - 50

1991 Tutti 1968 $20 - 20

1991 Tippee-Toes 1969 $12 - 18

1991 Magic Mindy 1970 $12 - 18

1991 Kopy Kate 1971 $12 - 18

1991 Dusty 1974/75 $12 - 18

1991 Donny & Marie 1977 $30 - 40

1992 Baby Cheerful Tearful 1968 $15 - 18

1992 4 Nursery Dolls 1959 $25 - 30

1992 Walt Disney's Cinderella 1965 $60 - 80

1992 Kiddle Kolognes 1969 $35 - 45   1992 Fashion Flatsy 1971 $8 - 12

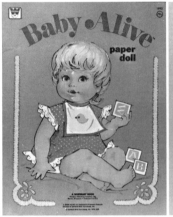
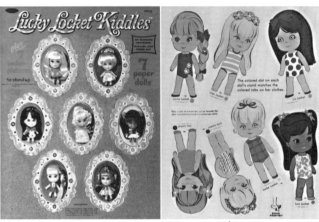

1992 Baby Alive 1973/75 $8 - 12   1993 Lucky Locket Kiddles 1967 $35 - 45

1993 One Hundred and One Dalmatians 1960 $50 - 75

1993 Cheerful Tearful 1966 $15 - 18   1993 Sweet-Treat Kiddles 1969 $35 - 45

1993 Finger Ding Paper Dolls 1971 $8 - 12   **1993** Ballerina Barbie 1977 $20 - 25

1994 Flatsy Paper Dolls 1970 $15 - 20   **1994** Malibu Barbie/The Sun Set 1972 $20 - 25

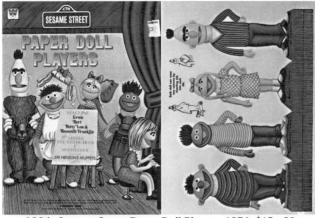

1994 Sesame Street Paper Doll Players 1976 $15 - 20   **1995** The Lennon Sisters 1963 $75 - 100

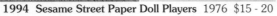

1995 Jill and Joan 1965 $25 - 30   **1995** Peg, Nan, Kay, Sue 1966 $25 - 30

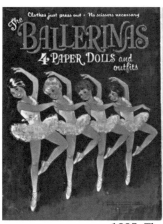
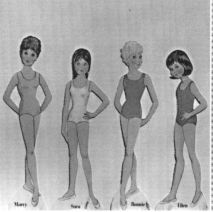

1995 The Ballerinas 1967 $25 - 30   **1995** Dolly Dears 1967 $25 - 30

1995 The Young Set 1973 $15 - 20   **1995** Calico Cathy 1976 $15 - 20

1995 The Waltons 1975 $35 - 50

1995-1 The Sunshine Fun Family 1978 $10 - 12

1996 Dennis the Menace
1960 $75 - 90

1996 Crissy 1970 $25 - 30

1996 Crissy and Velvet 1971 $25 - 30

1996 Barbie's Friend Ship 1973 $20 - 25

1996 Barbie's Beach Bus 1976 $20 - 25

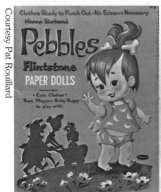

1997 Pebbles Flintstone 1963 $40 - 60

1997 Tammy and Her Family 1964 $50 - 75

1997 Tammy and Pepper 1965 $50 - 75

1997 Baby First Step 1965 $18 - 20

116

1997 Heidi, Hildy & Jan 1967 $20 - 25   **1997** Sketchy Double Playbook Fun 1970 $15 - 20

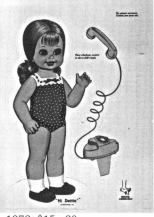

1997 "Hi Dottie" 1972 $15 - 20   **1997** Baby Thataway 1975 $10 - 15

1997-1 Fashion Photo Barbie and P.J.
1978 $20 - 25   **1998** Cathy Quick Curl 1975 $10 - 15

1999 Marge's Little Lulu
1960 $75 - 90

1999 Bedknobs and Broomsticks 1971 $30 - 45

1999 Twiggy 1967 $50 - 75   **1999** Raggedy Ann and Andy Circus Paper Dolls 1974 $15 - 20

1999 Baby Brother Tender Love 1977 $10 - 15   **2021** The Big Book of 10 Dolls and 100 Dresses 1934 $90 - 125

2042 Dinah Shore 1954 $100 - 150   **2043** Annie Oakley 1954 $75 - 90

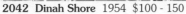

2044 Mary Hartline 1955 $100 - 175   **2048** Elaine Stewart 1955 $125 - 175

2049 Grace Kelly 1955 $125 - 175

2050 Pat Crowley 1955 $75 - 125

2054 Blondie 1955 $100 - 150 Published with six and eight pages.

2055 Jane Powell 1955 $125 - 175 Published with six and eight pages.

2056 Annie Oakley 1955 $75 - 90

2057 Elizabeth Taylor 1956 $125 - 200

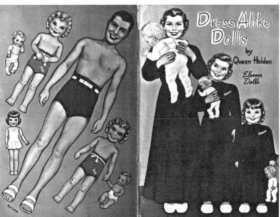

2053 Twin Dolls 1957 $25 - 35

2058 Dress Alike Dolls 1951 $75 - 125

119

2057 Elizabeth Taylor 1957 $125 - 200

2058 3 Little Girls Who Grew and Grew 1959 $45 - 60

2060 M-G-M Starlets 1951 $100 - 175 Front and back covers

2060 Dinah Shore 1956 $100 - 150

2061 Honey the Hair-Bow Doll 1950 $25 - 45

2061 Gale Storm 1958 $100 - 150

2062 Sunbonnet Sue
1951 $25 - 35

2062 Jr. Edition Fashion Show With Record 1973 $12 - 15

2067 Claire McCardell 1956 $65 - 80   2067 Bridal Fashion Show 1973 $15 - 20

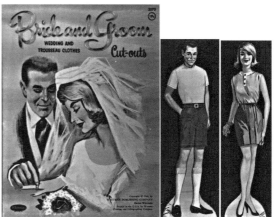

2069 Grace Kelly 1956 $125 - 175   2070 Bride and Groom 1963 $40 - 60

2071 Buttons and Billy 1963 $20 - 30   2071 Cute Quintuplets 1964 $40 - 50

2072 Baby Pat 1963 $25 - 30   2073 3 Pigs, Bears, Kittens Cut-Outs 1963 $40 - 50

2073 Twin Tots 1964 $20 - 30   **2074 Dolls of Other Lands** 1963 $20 - 30

2075 School Pals 1963 $20 - 30   **2081 Ricky Nelson** 1959 $75 - 100

2075 Slumber Party 1964 $25 - 35

2083 Walt Disney's Mouseketeer Annette 1958 $75 - 100   **2084 Cyd Charisse** 1956 $125 - 175

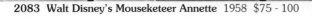

2084 4 Baby Dolls 1959 $35 - 45

2084 4 Playmates 1960 $20 - 30

2084 Birthday Party 1961 $40 - 60

2084 Prom Time 1962 $35 - 50

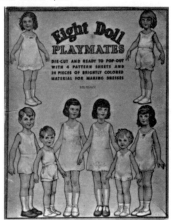

2085 Eight Doll Playmates $75 - 100
Box set, no date, listed in early 1930's catalog.

2085 Jane Powell 1957 $125 - 175

2085 Edd "Kookie" Byrnes 1959 $60 - 80

2085 Beautiful Bride 1960 $50 - 75

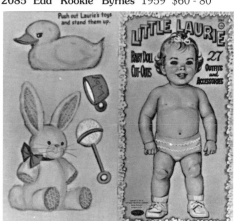

2085 Little Laurie 1961 $30 - 40

2086 Vera Miles 1957 $125 - 150   **2086 Natalie Wood** 1958 $100 - 175

2087 Rock Hudson 1957 $60 - 80   **2089 June Allyson** 1957 $125 - 175

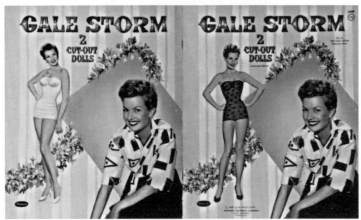

2089 Gale Storm 1959 $100 - 150   **2089 Carol Lynley** 1960 $60 - 80

2089 Children From Other Lands 1961 $20 - 30   **2089 Ginny Tiu** 1962 $30 - 40   **2090 A Dozen Cousins** 1960 $35 - 50

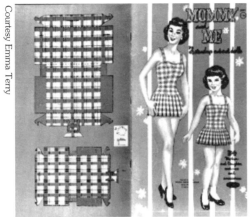

2091 Mommy and Me 1960 $25 - 35   **2091 Ballet** 1961 $25 - 35   **2091 Molly Bee** 1962 $50 - 75

2092 Playmates 1958 $25 - 30   **2093 Mary Ann, Mary Lou, Mary Jane** 1959 $50 - 75

2093 My Doll Melissa 1961 $25 - 30   **2099 Betty Hutton and her Girls** 1951 $100 - 150

2100 New Type Stay-On Clothes 1952 $35 - 45   **2101 I Love Lucy** 1953 $125 - 175

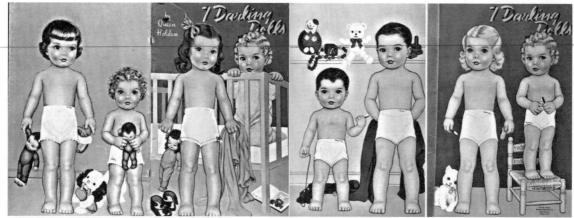

2102 7 Darling Dolls 1952 $50 - 75

2102 Here's the Bride 1954 $65 - 90

2103 Doris Day 1952 $125 - 175

2104 Mary Hartline 1952 $100 - 175

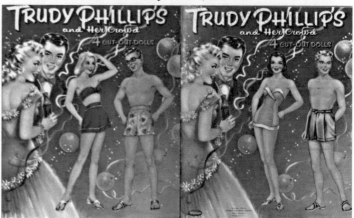

2104 Trudy Phillips and Her Crowd 1954 $60 - 80

2105 Sports Time
1952 $25 - 35

2105 Play Time
1952 $25 - 35

2105 Party Time
1952 $25 - 35

2105 Skating Stars 1954 $35 - 50

2106 Kit the 20" Doll 1952 $75 - 90

2107 Doris Day 1953 $125 - 175

2107 Peek-A-Boo 1955 $20 - 30

2108 Ava Gardner 1953 $125 - 200

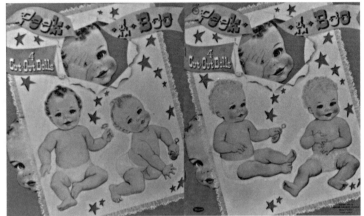

2112 Elizabeth Taylor 1954 $125 - 200

2109 Here's the Bride! 1953 $65 - 90

2116 Lucille Ball, Desi Arnaz with Little Ricky 1953 $125 - 175

2117 Lullaby Cut-Out Dolls 1953 $35 - 50

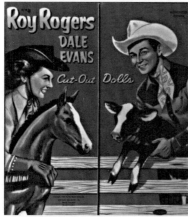

2118 Roy Rogers, Dale Evans 1953 $125 - 175

2119 Boys and Girls Doll Book 1955 $25 - 30

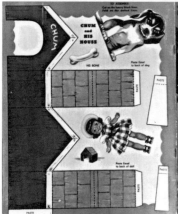

2129 Betty and Billy 1955 $45 - 60

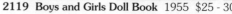

2131 Mommy and Me 1955 $30 - 40

Courtesy Virginia Crossley

2132 Tiny Tots 1956 $25 - 30

2133 Here's Trixie Belden 1956 $60 - 80

Courtesy Virginia Crossley

128

2139 Campus Queens 1957 $30 - 40

2145 3 Little Girls 1957 $20 - 30

2148 The Paper Doll Cut out Box 1939 $150 - 200 Contains 5 paper doll books labeled **#1002**.

2185 Walt Disney's Snow White and the Seven Dwarfs 1938 $200 - 300 Box set. The dwarfs are the same as those in **#970**, blue background book, except they are smaller (4-1/2" tall).

2190 The Dionne Quints 1937 $200 - 250 Dolls and clothes are from the five **#1055** books. The cars were not included.

2625 Dimple Triplets no date, 1950's $25 - 35

2627 Playmates no date, circa 1953 $25 - 35

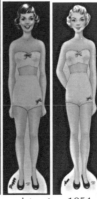

2627 Teen Time no date, 1950's $25 - 35

2627 Teen Time Dolls no date, circa 1954 $25 - 35

2630 Pat, A Fabrik Kit Doll no date, circa 1953 $25 - 35

2630 Terry, A Fabrik Kit Doll no date, circa 1953 $25 - 35

2945 2 Wood Dolls pictured in 1950 catalog $25 - 35

2966 3 Young Americans 1943 $65 - 85

2971 Sugarplum Dolls 1957 $25 - 35

2990 3 Wood Dolls Pictured in 1948 catalog $40 - 50

3005 Walt Disney's Snow White 1938 $125 - 175

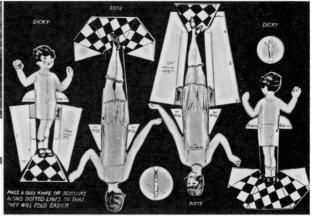

3013 Twelve Stand-Up Dolls – Stand-Up Toys no date, 1920's-30's $75 - 100

3036 Whopper Cut Out Doll Box $60 - 75 The dolls are from four different books (**#920, 965, 983 & 997**) and are reduced in size.

3046 Dolls of Many Lands 1931 $75 - 100 Showing five of the eight dolls

3046 Dolls of Many Lands 1932 $75 - 100 Showing eight of the ten dolls.

3059 4 Dolls 1933 $70 - 90 This is a reprint of **#920** with two redrawn dolls added. The clothes follow the same outlines as the original clothes, but have been redrawn. These same four dolls are also in the box set **#3082 Four Happy Dollies** 1933.

3063 Dolls and Dresses/The Big Little Set circa 1930's $70 - 90 Twenty dolls were reduced in size and taken from books **#917, 920** and **991**. They included crayons and 128 small sheets of clothes to be colored.

3081 10 Dolls from as Many Lands 1934 $90 - 125

131

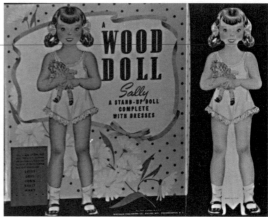

3945 American Nurse: A Wood Doll 1941 $35 - 50

3946 American Airlines Stewardess 1942 $35 - 50

3967 A Wood Doll: Sally 1941 $35 - 50

3967 A Wood Doll: Joan 1941 $35 - 50

3967 A Wood Doll: Mary 1941 $35 - 50

3967 A Wood Doll: Lois 1941 $35 - 50

3967 A Wood Doll: Betty 1941 $35 - 50

3980 Our WAVE Patsy 1943 $50 - 70

3980 Our WAAC Joan 1943 $50 - 70

3980 Our Nurse Nancy 1943 $50 - 70

3980 Our Soldier Jim 1943 $50 - 70

3980 Our Sailor Bob 1943 $50 - 70

3985 A Wood Doll: Babs no date, in 1947 catalog $30 - 40

3983 Big Big Doll
1943 $30 - 35

3984 Baby Dolls
1943 $30 - 35

Sets **3980**, **3983**, **3984**, and **3985** were sold in boxes or in envelopes with dolls glued to the front.

3985 A Wood Doll: Bunny
no date, in 1947 catalog $30 - 40

3985 A Wood Doll: Sissy
no date, in 1947 catalog $30 - 40

3985 A Wood Doll: This is Penny
no date, in 1947 catalog $30 - 40

3985 A Wood Doll: Ginger
no date, in 1947 catalog $30 - 40

4081 Big and Little Dolls 1954 $35 - 50

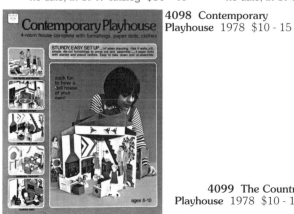

4098 Contemporary Playhouse 1978 $10 - 15

4099 The Country Playhouse 1978 $10 - 15

4100 A Wood Doll: This is Nancy
no date, in 1948 catalog $30 - 40

4100 A Wood Doll: This is Pat
no date, in 1948 catalog $30 - 40

4100 A Wood Doll: This is Susan
no date, in 1948 catalog $30 - 40

4100 A Wood Doll: This is Margie
no date, in 1950 catalog $25 - 35

4100 A Wood Doll: This is Betty
no date, in 1950 catalog $25 - 35

4100 A Wood Doll: This is Sandy
no date, in 1950 catalog $25 - 35

4105 Brother and Sister Dolls
no date, in 1949 catalog $25 - 35

4112 Peggy and Peter no date, in 1950 catalog $25 - 35

4116 A Wood Doll: Penny
no date, circa 1950's $25 - 35

4116 A Wood Doll: Sunny
no date, circa 1950's $25 - 35

4123 The Roly-Poly Twins no date, in 1952 catalog $25 - 35

4124 Penny: A "Sweetie" Doll
no date, circa 1953 $25 - 35

4124 Polly: A "Sweetie" Doll
no date, circa 1953 $25 - 35

4124 Sweetie Doll: Honey
no date, circa 1953 $25 - 35

4124 Sweetie Doll: Sugar
no date, circa 1953 $25 - 35

4125 The Polka Dot Tots no date, circa 1950's $25 - 35

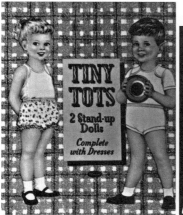

4129 Tiny Tots no date, circa 1954 $25 - 35

4129 Tiny Tots no date, circa 1956 $25 - 35

4129 Sonny and Sue no date, circa 1954 $25 - 35

4129 Gerry and Ginger no date, circa 1954 $25 - 35

4155 A Teeny Tiny Paper Doll 1967 $10 - 15

4195 This Is Lisa 1978 $9 - 12

4196 This Is Sue Ann 1978 $9 - 12

4197 This Is Marcie 1978 $9 - 12   **4198** This Is Katie 1978 $9 - 12   **4212** The Ginghams: Becky's School Room 1978 $15 - 20

4214 The Ginghams: Katie's Country Store 1978 $15 - 20   **4215** The Ginghams: Sarah's Farm 1978 $15 - 20

4216 The Ginghams: Carrie's Birthday Party 1978 $15 - 20   **4217** The Ginghams: Sarah's Picnic 1976 $15 - 20

4218 The Ginghams: Carrie's Bedroom 1976 $15 - 20   **4219** The Ginghams: Becky's Play Room 1976 $15 - 20

136

 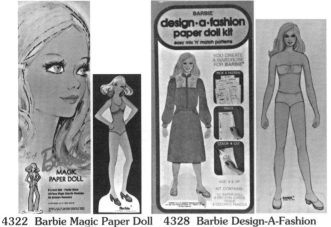   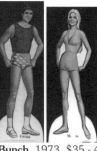

4220 The Ginghams: Katie's Ice Cream Parlor
1976 $15 - 20

4319 Raggedy Ann and Andy
1973 $15 - 20

4320 The Brady Bunch 1973 $35 - 45

4322 Barbie Magic Paper Doll
1973 $20 - 30

4328 Barbie Design-A-Fashion
1979 $12 - 18

4328-21 Barbie Design-A-Fashion
1982 $12 - 18

4329 Denim Deb's Design-A-Fashion
1979 $10 - 12

4329-21 Skipper Design-
A-Fashion 1982 $12 - 18

4330 Dawn
Magic Paper Doll
1971 $18 - 25

4331 Barbie Magic
Paper Doll
1971 $25 - 35

4332 New 'N'
Groovy P.J.
1971 $20 - 30

4333 Miss America
1974 $10 - 12

4334 The Waltons
1974 $30 - 40

4335 Baby Tender
Love 1974 $10 - 12

4336 Newport Barbie and Ken
1974 $20 - 30

4337 The Sunshine Family
1974 $10 - 12

4338 Sun Valley Barbie and
Ken 1974 $20 - 30

4339 Mrs. Beasley
1974 $15 - 18

Dolls same as **#4336**, clothes new.

137

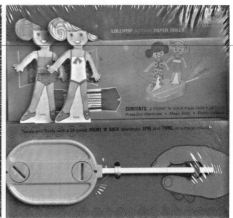

4340 Whirlikids Carousel Action
1971 $10 - 15

4341 Whirlikids Four Seasons Action
1971 $10 - 15

4342 Whirlikids Lollipop Action
1971 $10 - 15

4343 World of Barbie Play Fun Box 1972 $30 - 40

4375 Wedding Bell 1971 $35 - 50

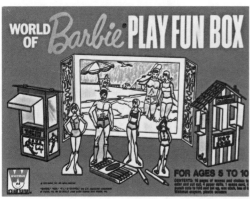

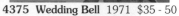

4347 Barbie Country Camper and Paper Dolls 1973 $30 - 45

4376 World of Barbie 1972 $35 - 50

4388 Shirley Temple
1976 $20 - 30

4389 Barbie and Ken All Sports
Tournament 1976 $20 - 30

4390 Pippi Longstocking 1976 $25 - 30

4391 Ballerina Barbie
1976 $25 - 35

4392 Barbie and Francie
1976 $25 - 35

4393 Malibu Francie
1976 $25 - 35

4394 Dusty
1974/76 $15 - 20

4395 Skipper
1976 $15 - 20

4396 Raggedy Ann and Andy
1975 $12 - 18

4397 Drowsy
1975 $12 - 18

4398 Baby Alive
1973/75 $12 - 18

4399 Quick Curl Barbie
1975 $25 - 30

4401 Teen Time no date, circa 1956 $25 - 30

4401 Teen Time Dolls 1958 $25 - 30

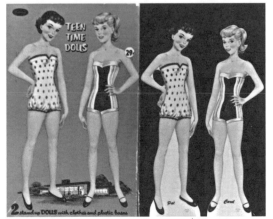

139

4401 Teen Time Dolls 1959 $25 - 30

4401 Teen Time Dolls 1960 $25 - 30

4401 Amy Magic Doll 1965 $15 - 20   **4401** Sunny Magic Doll 1966 $10 - 12   **4411** Big and Little Sister 1962 $15 - 20

 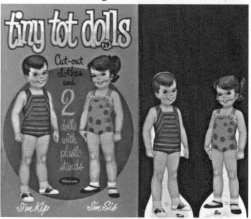

4412 Tiny Tots 1959 $15 - 25   **4412** Tiny Tot Dolls 1962 $15 - 20

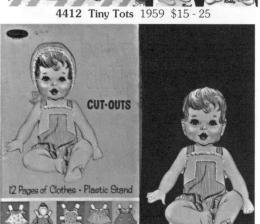 

4415 Baby Doll 1962 $15 - 25   **4415** Playmate 1965 $15 - 20

4574 Family of Dolls 1960 $35 - 50 Six dolls taken from five different sets.

4601 Barbie 1963 $65 - 85

4601 Heidi 1967 $20 - 30

4605 Barbie Wedding Dress'n
Fashion Clothes 1964 $65 - 85

4605 Barbie Fashion Window
Wardrobe 1965 $65 - 85

4607 Miss Teen Cut Outs 1960 $25 - 35 each
Mounted on cardboard with transparent wrap.

4607 Skipper Day-by-Day Wardrobe
1965 $65 - 85 Doll marked "1964."

4607 Skipper Fashion Calendar
Wardrobe 1965 $65 - 85

4607 Baby PeeWee
1968 $10 - 15

4609 Patty Duke
1965 $45 - 60

4610 Lucy
1963 $50 - 75

4612 National Velvet
1962 $45 - 60

4612 Bride and Groom
1966 $15 - 25

4612 Bridal Doll Box
1968 $15 - 25

4613 Janet Lennon
1962 $45 - 60

4613 Gretchen 1966 $20 - 25   4613 Trisha 1968 $20 - 25   4614 Connie Stevens
1961 $60 - 85

4616 Bunny Lou 1960 $20 - 30
Mounted on cardboard with
transparent wrap.

4616 Mary Lu 1961 $20 - 30
Mounted on cardboard with
transparent wrap.

4616 Barbie Travel Wardrobe
1964 $65 - 85

4618 Magic Stay-On Doll 1963 $20 - 25

4618 Sue the Magic Doll
1964 $20 - 25

4618 Vicki Paper Doll
1966 $20 - 25

4618 Amy Magic Doll
1968 $20 - 25

4620 Tammy School and Sports
Clothes 1964 $60 - 75

4620 Tammy A Closet Full of
Clothes 1965 $60 - 75

4621 Annette 1962 $60 - 75

4621 Mary Poppins
1966 $30 - 40

4622 Tutti 1967 $30 - 40

4624 Here's Mary no date, circa 1956 $35 - 45

4624 Karen Has Real Curls 1958 $30 - 35

4624 Peggy Has Real Curls
1957 $35 - 45

4624 Stephanie
1969 $12 - 15

4625 Bridal Party
no date, circa 1955 $50 - 75

4625 Bridal Party
no date, circa 1957 $50 - 75

4625 Bridal Party
1959 $50 - 75

4625 Baby Sister & Me
1969 $15 - 18

4626 Magic Stay-On Dresses for Patsy no date, circa 1955 $30 - 35

4626 Magic Stay-On Dresses Sally, no date, circa 1956 $30 - 35

4626 Magic Stay-On Dresses Susie, no date, circa 1957 $30 - 35

143

4626 Magic Stay-On Dresses Mary Anne, no date, circa 1958 $30 - 35

4626 Magic Stay-On Dresses Julie 1959 $30 - 35

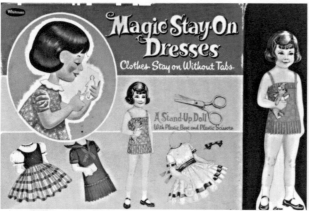

4626 Magic Stay-On Dresses Carol 1960 $30 - 35

4626 Tammy Date and Dress-
up Clothes 1964 $60 - 75

4626 Rag Doll 1969 $12 - 15

4636 Mickey and Minnie Mouse
1972 $20 - 35

4639 Skooter Fashion-Go-
Round 1965 $65 - 85

4640 Pepper Around-the-Clock
Wardrobe 1965 $60 - 75

4643 Baby's Hungry
1967 $12 - 15

4671 Mod Missy
1969 $12 - 15

4672 Eloise
1969 $12 - 15

4673 Daisy Darling
1969 $12 - 15

4683 Teddy Bear Sew-Ons
1978 $12 - 15

4684 Rag Doll Sew-Ons
1978 $12 - 15

4687 Sleeping Beauty 1967
$35 - 40 Some boxes not dated.

4701 Magic Stay-On Paper Dolls
Judy and Joan, 1963 $30 - 35

4701 Magic Stay-On 2 Paper Dolls
Sara & Pat, 1964 $30 - 35

4701 Magic Stay-On 2 Paper Dolls
Terry & Holly, 1966 $30 - 35

4701 Dolly Dears
1967 $12 - 15

4702 Mary Poppins 3 Magic Dolls
1964 $30 - 40

4704 Twiggy
1967 $25 - 30

4708 My Dolls Take a Trip
1957 $35 - 40

4712 Alice in Wonderland
1972 $40 - 50

4718 Lennon Sisters 1959 $75 - 100

4718 Lennon Sisters 1960 $75 - 100

4719 Liddle Kiddles Doll Box 1967 $60 - 80

4718 Malibu P.J.
1972 $20 - 30

4719 Ballet Paper Dolls 1968 $18 - 20

4719 Angelique
1971 $12 - 15

4721 Cinderella and the Prince
1972 $35 - 40

4721 Pepper and Dodi 1965 $50 - 75

4723 Sleeping Beauty Dolls 1958 $75 - 100

4722 Skediddle Kiddles 1968 $30 - 40

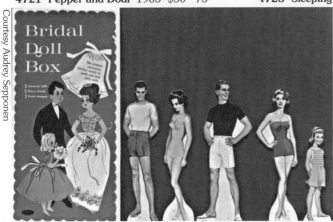

4723 Bridal Doll Box 1963 $50 - 60

4723 Bridal Doll Box 1964 $50 - 60

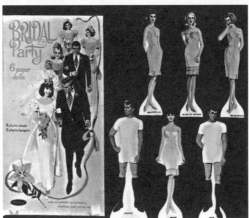

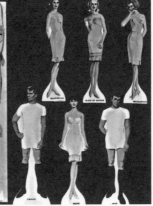

4723 Wedding Day Dolls 1961
$50 - 60 Dolls are the same as
#1969 Julie's Wedding, but
clothes are new.

4723 Bridal Party 1966 $50 - 60

4723 Happy Bridal Paper Dolls
1971 $15 - 18

4723 Bridal Party 1968 $40 - 50   **4732 Little Ballerina Doll Kit** 1961 $40 - 50   **4732 Snow White and The Prince**
1967 $35 - 40

4733 Walt Disney Starlets 1960 $85 - 125   **4733 Tammy and Her Family** 1964 $40 - 60   **4733 Growing Sally** 1969 $12 - 15
Dolls are from **#1997** but smaller, clothes are new.

4734 Sabrina   **4735 Barbie**   **4740 Cheerful Tearful**   **4740 Raggedy Ann and Andy**   **4744 Betsy McCall**
1971 $25 - 30   1971 $40 - 50   1966 $20 - 30   1968 $25 - 40   1971 $25 - 40

4743 PeeWee 1966 $35 - 45   **4743 Archie** 1969 $40 - 50   **4754 Winking Winny**   **4754 Crissy**
1969 $12 - 15   1972 $25 - 30

4755 Tippee Toes 1969 $12 - 18   **4756 Flatsy** 1969 $12 - 18

4758 Peepul Pals 1967 $12 - 18
Four dolls from **#1984**, clothes are new.

4760 Tippy Tumbles
1969 $12 - 18

4764 Magic Stay-On Dresses 1958 $35 - 45

4762 My Very Best Friend 1971 $12 - 15 **4763 Barbie** 1969 $45 - 60

4764 Buffy and Jody
1970 $35 - 50

4767 5 Family Affair Paper Dolls 1968 $35 - 50

4770 Little Lulu
1972 $35 - 45

Courtesy Audrey Sepponen

4770 Life Size Doll
1959 $25 - 35 Mounted on
cardboard with transparent wrap.

4772 Brigitte
1971 $10 - 12

4773 Green Acres
1968 $60 - 75

4774 Lucky Locket Kiddles
1968 $40 - 50

148

4774 Beautiful Crissy
1971 $25 - 30

4775 Patty Duke
1965 $35 - 45

4776 Chatty Baby
1963 $40 - 50

4777 Dollikin
1971 $10 - 12

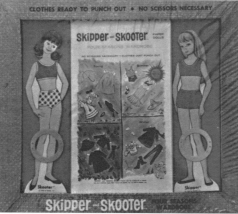

4777 Charmin' Chatty
1964 $25 - 35

**4778 Skipper and Skooter Four Seasons
Wardrobe** 1965 $75 - 90

4780 Swingy
1969 $20 - 25

4784 The Brady Bunch
1972 $35 - 45

4785 Barbie and Midge Travel Wardrobe
1965 $75 - 90

4785 Barbie
1967 $45 - 60

4786 Baby Nancy
1971 $15 - 18

4793 Barbie and Francie
1966 $50 - 75

4791 Pebbles and Bamm Bamm 1964 $35 - 50

4793 Barbie, Midge, Skipper 1965 $60 - 75

4795 This is Pennie
1962 $25 - 35

4796 The Flintstones
1962 $50 - 60

4796 Pebbles and Bamm Bamm
1965 $35 - 45

4797 Barbie and Ken Traveling Dolls 1962 $60 - 75

4797 Barbie and Ken
1963 $60 - 75

4798 Lennon Sisters
1962 $65 - 90

5310 Three Little Girls and How They Grew
1943/44 $75 - 90

5332 The Gang no date,
pictured in 1949 catalog $60 - 75

5336 Trim Dotty's Dresses
no date, pictured in 1949 catalog $25 - 35

5338 Make Mary's Clothes
1949 $25 - 35

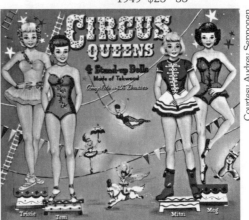

5340 Bridal Party Dolls 1949 $60 - 75

5368 Circus Queens 1957 $50 - 65

5347 **Outdoor Pals** no date, pictured in 1952 catalog $35 - 45

5349 **Bridal Party** no date, circa late 1940's-early 1950's $60 - 75

5349 **Bridal Party** no date, circa 1954 $60 - 75

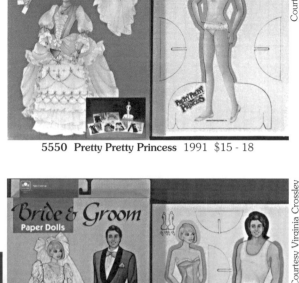

5550 **Pretty Pretty Princess** 1991 $15 - 18

5551 **Bride & Groom** 1991 $15 - 18

5552 **Barbie** 1991 $18 - 20

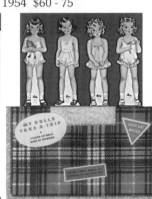

5608 **My Dolls Take A Trip**
(Set No. 1) circa 1950 $35 - 55

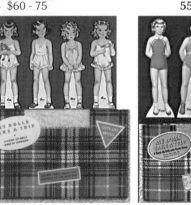

5608 **My Dolls Take A Trip Set No. 3** circa 1953 $35 - 50

5608 **My Dolls Take A Trip Set No. 4** circa 1955 $35 - 50

Set 2 is also **#5608**, same as No. 3, but suitcase is green plaid & some outfits are flocked in red.

5624 **Lucille Ball-Desi Arnaz and little Ricky** 1953 $90 - 125

6000 **Bride Originals Designer Set** 1991 $15 - 18

6001 **Star Originals Designer Set** 1991 $15 - 18

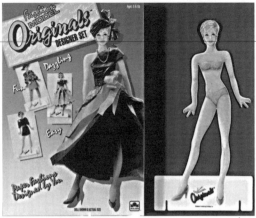

6002 Fashion Model Originals Designer Set 1991 $15 - 18

6901 New York World's Fair Goodwill Dolls 1939 $90 - 125

7316-A The Ginghams: Becky's Winter Carnival 1979 $15 - 20

7316-B The Ginghams, Katie At The Seashore 1979 $15 - 20

7316-C The Ginghams: Sarah's Country Kitchen 1979 $15 - 20

7316-D The Ginghams: Carrie's Boutique 1979 $15 - 20

7319-A The Ginghams: Becky's Tea Party 1981 $15 - 20

7319-B The Ginghams: Carrie's County Fair 1981 $15 - 20

7319-C The Ginghams: Katie's Artist Loft 1981 $15 - 20   **7319-D** The Ginghams: Sarah's Pet Shop 1981 $15 - 20

7407-A Poochie 1984 $10 - 12   **7407-B** Crystal Barbie 1984 $18 - 20   **7407-C** Rainbow Brite 1983/84 $12 - 14

7407-D Walt Disney's Donald and Daisy 1984 $12 - 15   **7408-A** Bride & Groom 1979 $15 - 18   **7408-A** Bride & Groom 1982 $12 - 15

7408A-1 Strawberry Sue 1981 $10 - 12   **7408-B** Fashion Photo Barbie 1979 $20 - 25   **7408-B** Walt Disney's Cinderella and the Prince 1982 $15 - 18   **7408B-1** Barbie 1981 $20 - 25   **7408-C** Denim Deb 1979 $10 - 12

7408-C Golden Dream Barbie
1982 $18 - 20

7408C-1 Super Teen Skipper & Scott
1981 $15 - 18

7408-D Snow White
1979 $18 - 20

7408-D Jean Jeans
1982 $10 - 12

7408D-1 Starr and Shawn 1981 $12 - 15

7408E Malibu Barbie 1982 $18 - 20

7408-F Walt Disney's Mickey & Minnie 1982 $12 - 15

7408-G Raggedy Ann and Andy 1982 $12 - 15

7408-H Western Barbie 1982 $18 - 20

7409-A Walt Disney's Donald & Daisy 1979 $15 - 18

7409-B Betsy Paper
Rag Doll 1979 $10 - 12

7409-C Teddy Bears
1979 $10 - 12

7409-D Trixie Belden
1979 $18 - 20

7410-A Ballet Stars
1980 $10 - 12

154

7410-B Pretty Changes Barbie 1980 $20 - 25

7410-C Star Princess and Pluta 1980 $12 - 15

7410-D Miss America 1980 $12 - 15

7411-A The Original Monchhichi Paper Dolls 1983 $10 - 12

7411-B Pink & Pretty Barbie 1983 $18 - 20

7411C Western Skipper 1983 $15 - 18

7411-D Barbie 1983 $18 - 20

7413-A Jean Jeans 1978 $10 - 12

7413-B Raggedy Ann and Andy 1978 $15 - 18

7413-C Rosebud 1978 $10 - 12

7413-D SuperStar Barbie 1978 $20 - 25

7554A-21 Amy 1981 $10 - 12

7554B-21 Jennifer 1981 $10 - 12

7554C-21 Jessica 1981 $10 - 12

7554D-21 Kristen 1981 $10 - 12

A7554E-22 **Beth** 1982 $10 - 12

A7554F-22 **Julie** 1982 $10 - 12

A7554G-22 **Peggy** 1982 $10 - 12

A7554H-22 **Rachel**
1982 $10 - 12

7820 **Calico Miss** 1975 $8 - 10
Small plastic-wrapped package

6853 **Little Brothers and Sisters** $75 - 90
This book is a reprint of four dolls from **#983**

PAPER DOLLS PUBLISHED BY WHITMAN

All original and reprint paper dolls published by the Whitman Publishing Company are identified in the list which starts below. If the paper dolls listed are reprints, the number of the original book will be in parentheses following the title. If a date is not known but the book was found in an old Whitman catalog, the date of the catalog is given. This does not necessarily mean the book was new that year, but it is a help in arriving at a date. Coloring books with paper dolls and books that contain paper toys or stand-up figures are included when known.

In 1983 the WHITMAN logo was dropped and the GOLDEN logo was adopted. Since the numbering system continued in the same manner, the new Golden paper doll books are included in this list.

6852 **Family of Paper Dolls** $75 - 90
This book is a reprint of the children dolls from #997

506	**Up To Date** two different sets (1) Bobby, Betty and Billie (2) Jack, Jane and Jerry. Circa 1924, pictured in 1920's catalog	
594	**I Like To Paint My Dolls** 1939, coloring/paper doll book	
657	**Scissor Cut-Outs** pictured in 1933 catalog, includes paper dolls	
674	**Judy Garland Fashion Paint Book** 1940, includes paper dolls	
682	**Fun and Play in Nimbletown** 1927 (includes paper dolls)	
826	**Play and Read** 1935 (includes paper dolls)	
900	**"Our Gang"** 1931	
900	**Peasant Costumes of Europe** 1934	
905	**Movie Stars Paper Dolls** 1931	
905	**Historic Costume Paper Doll Cut-Outs** 1934	
907	**Paper Doll "Joan", Paper Doll "Bobby"** 1928	
911	**Paper Doll "Gloria", Paper Doll "Sonny"** 1930	
915	**Here's Sally and Here's Sunny** 1939	
916	**Playtime Paint and Crayon Book** 1930's, paper toys	
917	**Nancy and Peggy** 1933	
917	**Jackie and Joan** 1933	
917	**Tommy and Jean** 1933	
917	**Davy and Dorothy** 1933	

917	**Polly and Peter** 1933
917	**Barbara and Baby Betty** 1933
920	**Baby Sister** 1929
920	**Baby Brother** 1929
920	**The Twins** 1932 (920)
920	**Baby Brother and Baby Sister** 1930 (920)
920	**Baby Brother and Baby Sister, The Twins** 1932 (920)
923	**A Book of Airplanes** 1930, stand-ups
925	**Ferdinand the Bull** 1938, stand-ups
928	**Alice in Wonderland** 1930, stand-ups
931	**Puppies and Kittens** 1939
932	**They Stand Up** 1936/39 (983)
933	**Aeroplane Cut Out Book** 1930, stand-ups
933	**Aeroplane Cut Out Book** listed in 1933 catalog, stand-ups
934	**Toys To Make** listed in 1930 catalog, stand-ups
934	**Playthings To Cut Out and Stand Up** 1930's, stand-ups
935	**The New Joan and Bobby** 1936
935	**Walt Disney's Pinocchio** 1939
935	**14 Dogs To Cut Out and Stand-Up** 1930, stand-ups
935	**22 Animals To Cut Out and Stand-Up** 1930, stand-ups
935	**19 Farm Yard Animals To Cut Out and Stand-Up** 1930, stand-ups

938	**Baby Nancy** 1931
938	**Little Orphan Annie, Mickey** 1931
938	**Little Orphan Annie** 1934
943	**Baby Ann** no date, circa 1932
945	**Valentines** 1935 (includes a paper doll)
946	**Raggedy Ann, Raggedy Andy** 1935
947	**Winnie the Pooh, Christopher Robin** 1935
948	**The Night Before Christmas with Cut-Outs** 1930's
949	**Christmas Tree Decorations To Make** 1936
950	**The Toy Shop** 1935, stand-up figures & furniture
950	**Walt Disney's Fantasia** 1940, stand-ups
951	**The Darlikin Dolls** 1938
952	**Christmas Cut-Outs** 1936
953	**The Circus Cut-Out Book** 1935, stand-ups
954	**Stencils Large and Small** 1930's, 30 punch-out animals
955	**Peter Rabbit** no date, pictured in 1939-40 catalog
956	**Paper Doll Gloria, Paper Doll Sonny** 1931 (911)
956	**Everything For Your Halloween Party** 1937, party items
957	**Flapper Fanny** 1938
958	**Life Size Paper Doll** 1936
959	**The Pop Out Book of The Three Bears, The**

1005	**Activity Fun** 1963, includes paper dolls
1010	**Betty Sue** 1939/41 (1055 Cecile)
1010	**Mary Jane** 1939/41 (1055 Emilie)
1010	**Mary Lee** 1939/41 (1055 Annette)
1010	**Patty Lou** 1939/41 (1055 Yvonne)
1010	**Sally Ann** 1939/41 (1055 Marie)
1011	**My Baby Book** 1942
1012	**Our Nurse Nancy** 1943 (3980)
1012	**Our Wave Joan** 1943 (3980 Patsy)
1012	**Mary of the Wacs** 1943 (3980 Joan)
1013	**Davy, Dorothy, Jackie and Joan** 1933 (combines two #917 books)
1013	**Nancy, Peggy, Tommy, Jean** 1933 (combines two #917 books)
1013	**Barbara, Betty, Polly, Peter** 1933 (combines two #917 books)
1013	**The Debutantes** Sandra & Sally, 1942 (#968 Four Mothers)
1013	**The Debutantes** June & Judy, 1942 (#968 Four Mothers)
1013-1	**Growing Up Skipper** 1978, coloring/paper doll book (1068)
1013-21	**Growing Up Skipper** 1978 (1068)
1015	**Betty and Joan** 1941/45 (3967)
1015	**Lois and Joan** 1941/45 (3967)
1015	**Mary and Joan** 1941/45 (3967)
1015	**Sally and Joan** 1941/45 (3967)
1015	**Barbie and Skipper** 1965/73, coloring book with puppy paper doll
1016	**Betty Brewer** 1942
1016	**Cora Sue Collins** 1942
1016	**Virginia Weidler** 1942
1017	**Baby Tender Love** 1971/76, coloring/paper doll book (1077)
1020	**Little Red Riding Hood Cut-Outs** 1939, stand-ups
1020	**Mother Goose Cut-Outs** 1939, stand-ups
1020	**The Three Bears Cut-Outs** 1939, stand-ups
1020	**The Three Little Pigs Cut-Outs** 1939, stand-ups
1025	**Play Dollies** 1920's (1040)
1033	**Mrs. Beasley** 1972, coloring/paper doll book
1035	**Roy Rogers' Double-R-Bar Ranch** 1955, coloring/paper doll book
1040	**Play Dollies** 1920's
1043	**Baby Beans** 1973, coloring/paper doll book
1043-1	**Here Comes The Bride** 1979, coloring/paper doll book (dolls from #1986 Bridal Doll Book)
1050	**Paper Doll** 1932 (911)
1050	**Baby Brother and Baby Sister** 1932 (920)
1050	**The Twins** 1932 (920)
1053	**Nancy and Sluggo** 1972, coloring/paper doll book
1054	**Eskimo Cut-Outs** 1938, stand-ups
1054	**Farm Cut-Outs** 1938, stand-ups
1054	**Indian Cut-Outs** 1938, stand-ups
1054	**Mexican Cut-Outs** 1938, stand-ups
1054	**Zoo Cut-Outs** 1938, stand-ups
1055	**Annette** 1936
1055	**Cecile** 1936
1055	**Emilie** 1936
1055	**Marie** 1936
1055	**Yvonne** 1936
1056	**Tiny Toddles** 1970, coloring/paper doll book
1060	**Baby Nancy** 1935, reduced in size (938)
1061	**Sweet Swingers** 1971, coloring/paper doll book
1061-31	**Here Comes The Bride** 1979, coloring/paper doll book (1043-1)
1068	**See America** 1971, activity book with paper dolls
1068	**Growing Up Skipper** 1978, coloring/paper doll book
1071	**Barbie and P.J.** 1973, coloring/paper doll book
1077	**Baby Tenderlove** 1971, coloring/paper doll book
1085	**Tubsy** 1968, coloring/paper doll book
1087	**Baby's Hungry!** 1968, coloring/paper doll book
1088	**8 Dolls, Over 100 Dresses** 1939 (990 Boys and Girls at School)

1094	**Francie** 1967, coloring/paper doll book
1105-22	**Little Dancers** 1972, coloring/paper doll book, dolls from #1968
1112	**Busy Book** 1971, includes paper dolls to color
1121	**Little Dancers** 1972, coloring/paper doll book (dolls from #1968)
1124	**Santa's Christmas Corner** 1983, activity book with one Santa paper doll
1130	**Bridal Book** 1968, coloring/paper doll book
1131	**Little Dancers** 1972, coloring/paper doll book (dolls from #1968)
1133	**Carol Heiss** 1961, coloring/paper doll book
1136-3	**Winnie the Pooh** 1978, activity book with paper dolls (1717)
1137	**Laurel and Hardy** 1968, coloring/paper doll book
1138	**Secret Sue** 1967, coloring/paper doll book
1139	**Bride and Groom** 1972, coloring/paper doll book
1140	**My Busy Book** 1963/72, coloring/paper doll book (3082-D)
1140-1	**Bugs Bunny & Friends Fun Book** 1979, with paper dolls (1352-31)
1141	**Patty Duke** 1960, coloring/paper doll book
1141	**Chatty Baby** 1961/62/63, coloring/paper doll book
1143-25	**Masters of the Universe** 1985, activity book with some paper dolls
1146	**Ann and Arthur** 1939 (990)
1146	**Cynthia and Bobby** 1939 (990)
1146	**Janet and Donny** 1939 (990)
1146	**Jayne and Tommy** 1939 (990)
1146	**Judy and Dick** 1939 (990)
1146	**Kitty and Billy** 1939 (990)
1146	**Laura and Jimmy** 1939 (990)
1146	**Muriel and David** 1939 (990)
1146	**Nancy and Tommy** 1939 (990)
1146	**Patsy and Donny** 1939 (990)
1146	**Robin and Jack** 1939 (990)
1146	**Susan and Johnny** 1939 (990)
1146-18	**Jewel Secrets Barbie** 1987, coloring/paper doll book (1146-29)
1141-29	**Jewel Secrets Barbie** 1987, coloring/paper doll book
1158	**Whopper Paint Book** includes a paper doll
1167	**Francie** 1967, coloring/paper doll book (1094)
1169	**Francie** 1967, coloring/paper doll book (1094)
1170	**All Size Dolls** 1945 (982)
1170	**Portrait Girls** 1947 (966)
1171	**Statuette Dolls Polly & Pat** 1946 (985)
1171	**Jane Powell** 1952
1171	**Jane Powell** 1953
1171	**My Pet Monster** 1986, with paper doll
1172	**Baby Dolls** 1950
1172	**Roy Rogers and Dale Evans** 1952
1172-1	**Jem** 1986, coloring/activity book with paper dolls (5500)
1173	**The Cradle Crowd** 1948
1173	**June Allyson** 1953
1173-1	**Jem** 1986, activity book with 4 paper dolls to color
1174	**Three Sisters** 1942
1174	**Blondie** 1953
1175	**Sandra and Sue** 1948 (1180)
1175	**Mary Hartline** 1953
1176	**Three Little Girls and How They Grew and Grew** 1945 (990)
1176	**Nursery School Dolls** 1953
1177	**Dress Me!** 1943 (970)
1177	**Elizabeth Taylor** 1953
1178	**Cloth-Like Clothes For 3 Cute Girls** 1949
1178	**Debbie Reynolds** 1953
1179	**Baby Cut-out Dolls** 1949
1179	**Doris Day** 1954
1180	**Sandra and Sue** 1948
1180	**Blondie** 1954
1181	**Sally** (Bride) 1950
1181	**We're A Family** 1954
1182	**Brother and Sister Statuette Dolls** 1950
1183	**Hopalong Cassidy** 1950, stand-ups
1183	**Barbie and Ken** 1962, coloring/paper doll book

1183	**Bridal Book** 1968, coloring/paper doll book (1130)
1184	**Jimmy and Jane visit Gene Autry at Melody Ranch** 1951
1185	**Jane Powell** 1951
1186	**Roy Rogers and Dale Evans** 1950
1187	**Bridal Party** 1950
1188	**Make Judy Laugh** 1952
1189	**Here Comes the Bride** 1952
1190	**June Allyson** 1950/52 (970)
1191	**Blondie** 1949/52 (reprint of #963 with two extra pages of clothes)
1192	**Ava Gardner** 1949/52 (reprint of #965 with two extra pages of clothes)
1193	**Elizabeth Taylor** 1950/52 (reprint of #973 with back & front cover dolls switched & swimsuit changed from blue to yellow)
1195	**40 Valentines** 1953
1206	**Winnie the Pooh** 1978, activity book with paper dolls (1717)
1218	**Jewel Secrets Barbie** 1987, coloring/paper doll book (1146-29)
1219	**Toys to Color** 1957, includes a paper doll
1250	**36 Christmas Cards** 1951
1286	**Easy To Do** 1978, activity book with paper dolls
1287	**A Bunny Fun Book** 1975, activity book with paper dolls
1287	**See It – Do It** 1978, activity book with paper dolls
1345-30	**Bunch O' Fun Activity Book** 1981, with paper dolls
1352-31	**Bugs Bunny and Friends Fun Book** 1979, activity book with paper dolls
1355	**Little Kiddles Press Out** stand-ups (date n/a)
1358	**Pretty Belles** 1965 (1966)
1359	**Sesame Street My Favorite Things** 1979, with paper dolls
1361	**Sesame Street Neighborhood** 1976, coloring/paper doll book
1386	**Huey, Dewey and Louie Super Dooper** with cut-outs 1978
1398	**Wonder Woman** 1979, coloring/paper doll book
1398-2	**Wonder Woman** 1979, coloring/paper doll book (1398)
1410	**Busy Book, Things To Do** 1971, activity book with paper dolls
1418	**Fun With Paper** 1974, paper toys
1426	**Wendy's Wardrobe Stencils** 1965
1426-21	**Busy Book Things to Do** 1971 (1410)
1437-1	**Preschool Read-Color-Play** 1971, activity book with paper dolls
1456	**Read-Color-Play** 1954/63 (4027)
1500	**Perfume Pretty Barbie** 1988
1501	**Bride & Groom** 1988
1502	**Barbie** 1990
1502-1	**Barbie** 1990
1502-2	**Barbie** 1990/91 (1502)
1502-3	**Barbie** 1992
1502-4	**Barbie** 1993
1502-5	**Barbie** 1990/93 (1502-1)
1503	**Lil Miss 'n Me** 1990
1506	**Quints** 1990
1521	**Barbie Day-To-Night** 1985 (1982-48)
1522	**Great Shape Barbie** 1985 (1982-49)
1523	**Tropical Barbie** 1986
1523-1	**Tropical Barbie** 1986 (1523)
1523-2	**Barbie** 1990
1524	**Jem** 1986
1525	**Peaches 'N Cream Barbie** 1985 (1983-48)
1526	**Hart Family** 1985 (1983-50)
1526-1	**Heart Family** 1985 (1983-50)
1527	**Barbie and Ken** 1984 (1985-51)
1528	**Barbie and the Rockers** 1986
1528-1	**Barbie and the Rockers** 1986 (1528)
1529	**Princess of Power** 1985 (1984-54)
1530	**Princess Diana** 1985 (1985-50)
1532	**Punky Brewster** 1986
1532-1	**Punky Brewster** 1986 (1532)
1533	**Sesame Street Paper Doll Seasons** 1984 (1985-48)

	Seven Dwarfs 1972 (1987)	
1980	The Sunshine Family 1977	
1980-1	The Ginghams 1976 (1985)	
1980-2	Baby This 'N That 1979	
1980-3	Super Teen Skipper 1980	
1980-22	Baby This 'N That 1979 (1980-2)	
1981	Walt Disney's Sleeping Beauty 1959	
1981	Walt Disney's Babes in Toyland 1961, large book of stand-ups	
1981	Walt Disney's It's a Small World 1966	
1981	Liddle Kiddles 1967	
1981	Storybook Kiddles 1968	
1981	New 'N' Groovy P.J. 1970	
1981	P.J. Cover Girl Paper Doll 1971	
1981	Barbie's Sweet 16 1974	
1981	Barbie and Her Friends, All Sports Tournament 1975	
1981-1	The Sunshine Family 1977 (1980)	
1981-2	Paper Doll Dancers 1979	
1981-3	Simply Sam 1980	
1982	The Flintstones 1961, large stand-ups of Fred, Barney, Wilma & Betty with outfits	
1982	Walt Disney's Mary Poppins 1964	
1982	Walt Disney Presents Mary Poppins 1966	
1982	Chitty Chitty Bang Bang 1968	
1982	Babykins 1970	
1982	Josie and the Pussycats 1971	
1982	Francie 1973, 1st issued folder style, then book	
1982	Baby Dreams 1976	
1982-1	Rosebud 1978	
1982-2	Raggedy Ann 1970 (1977)	
1982-21	Rosebud 1978 (1982-1)	
1982-23	Teddy Bear Family 1980	
1982-31	Starr 1980	
1982-32	Fashion Photo Barbie 1978 (1997-1)	
1982-33	Super Teen Skipper 1980 (1980-3)	
1982-34	Pretty Changes Barbie 1981	
1982-41	Starr 1980 (1982-31)	
1982-42	Pretty Changes Barbie 1981 (1982-34)	
1982-43	Western Barbie 1982	
1982-44	Sunsational Malibu Barbie 1983	
1982-45	Angel Face Barbie 1983	
1982-46	Twirly Curls Barbie 1983	
1982-47	Barbie "Fantasy" 1984	
1982-48	Day-to-Night Barbie 1985	
1982-49	Great Shape Barbie 1985	
1983	The McGuire Sisters 1959	
1983	Jill and Bill 1960	
1983	The Lennon Sisters 1961	
1983	The Jetsons 1963	
1983	Pebbles and Bamm-Bamm 1964	
1983	Pebbles and Bamm-Bamm 1965	
1983	Pebbles Punch Out 1965, stand-ups	
1983	Jill and Joan 1965 (1995)	
1983	Pebbles and Bamm-Bamm 1966	
1983	Tiny Tots 1967 (1977)	
1983	Tini-Mods, 6 Paper Dolls 1968 (1988)	
1983	Strawberry Sue 1973	
1983	Paper Dolls of Early America 1975	
1983	SuperStar Barbie 1977	
1983-1	Bridal Doll Book 1978 (1986)	
1983-2	SuperStar Barbie 1977 (1983)	
1983-31	Strawberry Sue 1979 (1976-2)	
1983-32	Simply Sam 1980 (1981-3)	
1983-33	My Best Friend 1980 (1978-3)	
1983-34	Freckles & Sniffles 1981	
1983-35	We're Twins 1981	
1983-41	Strawberry Sue 1979 (1976-2)	
1983-42	The Original Monchhichi 1982	
1983-43	Golden Dream Barbie 1982	
1983-44	Pink & Pretty Barbie 1983	
1983-45	Poochie 1983	
1983-46	Crystal Barbie 1984	
1983-48	Peaches 'n Cream Barbie 1985	
1983-50	The Heart Family 1985	
1984	The Happy Family 1960	
1984	Skipper 1965	
1984	Peepul Pals 1967	
1984	Raggedy Ann Paper Doll 1970 (1977)	
1984	Sleeping Beauty 1970 (1981)	
1984	Timey Tell 1971	
1984	Quick Curl Barbie 1973	

1984-31	Story-Book Beans 1980 (1979-3)	
1984-33	Rosebud 1978 (1982-1)	
1984-41	Story-Book Beans 1980 (1979-3)	
1984-42	We're Twins 1981 (1983-35)	
1984-43	Strawberry Sue 1979 (1976-2)	
1984-44	Butterscotch 1983	
1984-46	Freckles & Sniffles 1981 (1983-34)	
1984-46	Rainbow 1983	
1984-47	The Ginghams 1981 (1987-33)	
1984-48	Wedding 1984	
1984-50	My Very First Paper Doll 1983 (1985-46)	
1984-51	Rainbow Brite 1984	
1984-53	Star Fairies 1985	
1984-54	Princess of Power 1985	
1984-55	Sesame Street Presents Follow That Bird The Movie 1985	
1985	My Doll Family 1955	
1985	My Doll House Family, The Browns Live Here 1957	
1985	Tiny Chatty Twins 1963	
1985	Charmin' Chatty 1964	
1985	Scooter 1965	
1985	Buffy 1968	
1985	Buffy 1969	
1985	Star Charmers 1970	
1985	Barbie and Ken 1970 (1976)	
1985	Raggedy Ann 1970 (1977)	
1985	Raggedy Ann and Andy, Lolly-Pop Fashions 1972 (dolls from #1979)	
1985	The Ginghams 1976	
1985-31	Bridal Doll Book 1978 (1986)	
1985-32	Neighborhood Kids 1980 (1978-24)	
1985-33	Teddy Bear Family 1980 (1982-23)	
1985-34	Peek-A-Boo Baby 1981	
1985-41	Bridal Doll Book 1978 (1986)	
1985-42	Neighborhood Kids 1980 (1978-24)	
1985-43	Teddy Bear Family 1980 (1982-23)	
1985-45	Bugs Bunny & Honey Bunny 1983	
1985-46	My Very First Paper Doll 1983	
1985-47	My Little Sister 1983	
1985-48	Sesame Street Paper Doll Seasons 1984	
1985-49	Walt Disney's Mickey & Minnie 1983	
1985-50	Princess Diana 1985	
1985-51	Barbie & Ken 1984	
1985-60	Walt Disney Pictures/Return To Oz 1985	
1986	Walt Disney's Let's Build Disneyland 1957, stand-ups	
1986	Donald Duck and His Friends 1961, large stand-up figures of Donald & nephews, Goofy, Mickey Mouse & Pluto	
1986	Drowsy 1965	
1986	Lovable Babies 1966 (1978)	
1986	Francie and Casey 1967	
1986	Barbie and Ken 1970 (1976)	
1986	Playtime Pals 1970	
1986	Mrs. Beasley 1972	
1986	Shirley Temple 1976	
1986	Bridal Doll Book 1978	
1986-31	Walt Disney's Mickey & Minnie Steppin' Out 1977 (1979)	
1986-32	Walt Disney's Daisy and Donald 1978 (1990-1)	
1986-33	Walt Disney's Winnie-The-Pooh 1980 (1977-24)	
1986-41	Walt Disney's Mickey & Minnie Steppin' Out 1977 (1979)	
1986-43	Walt Disney's Winnie-The-Pooh 1980 (1977-24)	
1987	Sleeping Beauty Playtime Set 1959, stand-ups	
1987	Walt Disney presents Snow White and the Seven Dwarfs not dated, but 1967	
1987	The Archies 1969	
1987	World of Barbie 1971, six paper dolls	
1987	Casual Co. 1972	
1987	Little Lulu and Tubby 1974	
1987	Wizard of Oz 1976	
1987-1	The Archies 1969 (1987)	
1987-31	Trixie Belden with Honey Wheeler 1981	
1987-32	Raggedy Ann and Andy 1980 (1977-23)	
1987-33	The Ginghams Visit Grandma 1981	
1987-34	The Ginghams 1976 (1985)	

1987-41	Trixie Belden with Honey Wheeler 1981 (1987-31)	
1987-42	The Ginghams Visit Grandma 1981 (1987-33)	
1987-43	Raggedy Ann and Andy 1980 (1977-23)	
1988	The Wedding Playbook 1960	
1988	Winnie's Wardrobe Doll Book 1966	
1988	Baby Go-Along 1968	
1988	Tini Mods 1968	
1988	Baby Go Bye-Bye 1971	
1988	Newborn Baby Tender Love 1973	
1988	Big Jim and Big Jack 1976	
1988-1	Raggedy Ann and Andy 1978	
1989	Santa's Workshop 1960, includes paper dolls	
1989	Santa's Toyland Playbook 1962, includes paper dolls	
1989	Wedding Paper Dolls 1966	
1989	Bride and Groom 1970	
1989	Bride Doll Book 1972	
1989	Bridal Fashions for Five Weddings 1973	
1989	Barbie Fashion Originals 1976	
1990	The Nutcracker Ballet Cutouts 1960	
1990	Santa's Playbook 1964, includes paper dolls	
1990	Wedding Paper Dolls 1966 (1989)	
1990	Missy Go-Along 1970	
1990	Barbie Country Camper 1973	
1990	Growing Up Skipper 1976	
1990-1	Walt Disney's Daisy and Donald 1978	
1990-21	Walt Disney's Daisy and Donald 1978 (1990-1)	
1991	Lennon Sisters 1959, folder color is green or pink, both sets identical	
1991	4 Nursery Dolls 1959 (1992)	
1991	Dennis the Menace Back-Yard Picnic 1960	
1991	Lucy and Her TV Family 1963	
1991	Patty Duke 1964	
1991	Patty Duke 1965	
1991	Tutti 1968	
1991	Tippee-Toes 1969	
1991	Magic Mindy 1970	
1991	Kopy Kate 1971	
1991	Dusty 1974/75	
1991	Donny & Marie 1977	
1992	4 Nursery Dolls 1959	
1992	Walt Disney's Cinderella 1965	
1992	Lucky Locket Kiddles 1967 (1993)	
1992	Baby Cheerful Tearful 1968	
1992	Kiddle Kolognes 1969	
1992	Fashion Flatsy 1971	
1992	Baby Alive 1973/75	
1993	One Hundred and One Dalmatians 1960	
1993	Cheerful Tearful 1966	
1993	Lucky Locket Kiddles 1967	
1993	Sweet-Treat Kiddles 1969	
1993	Finger Ding Paper Dolls 1971	
1993	Mrs. Beasley 1972 (1986)	
1993	Ballerina Barbie 1977	
1993-1	Ballerina Barbie 1977 (1993)	
1994	Santa 'N Tree 1961, 20" Santa & other stand-ups	
1994	The Archies 1969 (1987)	
1994	Flatsy Paper Dolls 1970	
1994	Malibu Barbie, The Sun Set 1972	
1994	Sesame Street Paper Doll Players 1976	
1995	The Lennon Sisters 1963	
1995	Jill and Joan 1965	
1995	Peg, Nan, Kay, Sue 1966	
1995	Dolly Dears 1967	
1995	The Ballerinas 1967	
1995	Buffy 1968 (1985)	
1995	The Young Set 1973	
1995	The Waltons 1975	
1995	Calico Cathy 1976	
1995-1	The Sunshine Fun Family 1978	
1995-21	The Sunshine Fun Family 1978 (1995-1)	
1996	Dennis the Menace 1960, large stand-ups with few outfits	
1996	Barbie Has A New Look 1967 (1976)	
1996	Crissy 1970	
1996	Crissy and Velvet 1971	
1996	Barbie's Friend Ship 1973	
1996	Barbie's Beach Bus 1976	

1996-1	Barbie's Beach Bus 1976 (1996)
1997	Pebbles Flintstone 1963
1997	Tammy and Her Family 1964
1997	Tammy and Pepper 1965
1997	Baby First Step 1965
1997	Heidi, Hildy & Jan 1967
1997	Tini Mods 1968 (1988)
1997	Sketchy Double Playbook Fun 1970
1997	"Hi Dottie" 1972
1997	The Brady Bunch 1973 (1976)
1997	Baby Thataway 1975
1997-1	Fashion Photo Barbie and P.J. 1978
1997-21	Fashion Photo Barbie and P.J. 1978 (1997-1)
1998	Yogi Bear, Huckleberry Hound 1960, very large stand-ups, some hats & ties
1998	The Ballerinas 1967 (1995)
1998	Walt Disney Presents Snow White and the Seven Dwarfs (1987)
1998	Cathy Quick Curl 1975
1999	Marge's Little Lulu Giant Stand-Up Cut-Out 1960
1999	Twiggy 1967
1999	Bedknobs and Broomsticks (Disney) 1971
1999	Raggedy Ann and Andy Circus Paper Dolls 1974, 69¢ book has 6 pages of clothes, 79¢ book has 4 pages of clothes
1999	Baby Brother Tender Love 1977
2000	Let's Shop 1994
2001	Prom Night 1991 (1688)
2002	Muppet Babies 1991
2003	Miss Piggy 1994
2004	Kiddieland Village no date, 1930's, stand-ups
2004	Walt Disney's Minnie Mouse 1990/94 (1692)
2005	Blossom 1994
2006	The Little Mermaid 1992/94 (1684)
2007	Snow White 1991/94 (1697)
2007-01	Snow White no date (1697)
2008	Cinderella 1989/94 (1545)
2008-30	Cinderella no date (1545)
2008-31	Cinderella no date (1545)
2010	Disney's Beauty & The Beast 1991/94 (1675)
2011	Walt Disney's Sleeping Beauty 1994
2011-01	Walt Disney's Sleeping Beauty (2011)
2013-14	Blossom 1994 (2005)
2015	Aladdin 1992/94 (1606)
2016	Little Mermaid Film Fashions 1997
2018	Barbie 1994
2019	Anastasia 1997
2021	The Big Book of 10 Dolls and 100 Dresses 1934
2042	Dinah Shore 1954
2043	Annie Oakley 1954
2044	Mary Hartline, TV's Golden Princess 1955
2048	Elaine Stewart 1955
2049	Grace Kelly 1955
2050	Pat Crowley 1955
2050	Santa's Magic Tree, Scented Gingerbread 'N' Pine Cone Ornaments 1971
2051	Playtime Pop-Outs 1972, paper toys
2053	4 Baby Dolls 1958, dolls from 1173 but smaller, clothes new
2053	Twin Dolls 1957
2054	Blondie 1955 (6 & 8 page books)
2055	Jane Powell 1955 (6 & 8 page books)
2056	Annie Oakley with Tagg and Lofty 1955
2057	Elizabeth Taylor 1956
2057	Elizabeth Taylor 1957
2058	Dress Alike Dolls 1951
2058	3 Little Girls Who Grew and Grew 1959
2060	4 Sleeping Dolls 1945 (977 & 978)
2060	M-G-M Starlets 1951
2060	Dinah Shore 1956
2061	Honey the Hair-Bow Doll 1950
2061	Gale Storm 1958
2061	Bridal Fashion Show With Record 1973 (2067)
2062	Sunbonnet Sue 1951
2062	Jr. Edition Fashion Show With Record 1973
2063	Lot's of Fun 1951, activity book with "Dancing Doll paper doll"
2063	Busy Airport 1973, stand-ups & record (2069)

2064	Funtime Circus 1973, stand-ups & record (2070)
2066	40 Valentines 1951
2067	Claire McCardell 1956
2067	Bridal Fashion Show With Record 1973
2068	Jr. Edition Fashion Show With Record 1973 (2062)
2069	Grace Kelly 1956
2069	Busy Airport 1973, stand-ups & record
2070	Bride and Groom Wedding 1963
2070	Funtime Circus 1973, stand-ups & record
2071	Buttons and Billy 1963
2071	Cute Quintuplets 1964
2072	Baby Pat 1963
2073	3 Pigs, Bears, Kittens Cut-Outs 1963
2073	Twin Tots 1964
2073	Bridal Fashion Show With Record 1973 (2067)
2073	Santa's Helper Gingerbread Village 1976, stand-ups
2074	Dolls of Other Lands 1963
2074	Jr. Edition Fashion Show With Record 1973 (2062)
2075	School Pals 1963
2075	Slumber Party 1964
2075	Busy Airport 1973, stand-ups & record (2069)
2076	Funtime Circus 1973, stand-ups & record (2070)
2078	Simple Objects to Push Out and Put Together 1960, paper toys
2079	Run the Farm With Equipment 1960, stand-ups
2079	Bridal Fashion Show With Record 1973 (2067)
2080	Road Builders to Push Out and Assemble 1960, stand-ups
2080	Jr. Edition Fashion Show With Record 1973 (2062)
2081	Ricky Nelson 1959
2081	Busy Airport 1973, stand-ups & record (2069)
2082	Circus Stage Show to Push Out and Put Together 1960, stand-ups
2082	Funtime Circus 1973, stand-ups & record (2070)
2083	Walt Disney's Mouseketeer Annette 1958
2084	Cyd Charisse 1956
2084	4 Baby Dolls 1959
2084	4 Playmates 1960
2084	Birthday Party 1961
2084	Prom Time 1962
2085	Eight Doll Playmates box set, in early 1930's catalog
2085	Jane Powell 1957
2085	Edd "Kookie" Byrnes 1959
2085	Beautiful Bride 1960
2085	Little Laurie 1961
2085	Ballet 1961 (2091)
2086	Vera Miles 1957
2086	Natalie Wood 1958
2087	Rock Hudson 1957
2088	Elizabeth Taylor 1957 (2057)
2088	Debbie Reynolds 1955/59, #1955 for dolls, clothes are different
2089	Jingle Bells 1950, novelty cut-outs & coloring book
2089	June Allyson 1957
2089	Gale Storm 1959
2089	Carol Lynley 1959
2089	Children From Other Lands 1961
2089	Ginny Tiu 1962
2090	Sports Time 1952 (2105)
2090	A Dozen Cousins 1960
2090	Children From Other Lands 1961 (2089)
2091	Party Time 1952 (2105)
2091	Mommy and Me 1960
2091	Ballet 1961
2091	Little Laurie 1961 (2085)
2091	Molly Bee 1962
2092	Playmates 1958
2093	Mary Ann, Mary Lou, Mary Jane 1959
2093	Mommy and Me 1960 (2091)
2093	My Doll Melissa 1961
2094	Crazy Racers 1960, stand-ups

2099	Four Cut-Out Paper Dolls box set (907 & 911)
2099	Betty Hutton and her Girls 1951
2099	Trim A Tree 1968, ornaments to make
2100	New Type Stay-On Clothes 1952
2101	I Love Lucy 1953
2102	7 Darling Dolls 1952
2102	Here's the Bride 1954
2103	Doris Day 1952
2104	Mary Hartline 1952
2104	Trudy Phillips and Her Crowd 1954
2105	Sports Time 1952
2105	Play Time 1952
2105	Party Time 1952
2105	Skating Stars 1954
2106	Kit the 20" Doll 1952
2107	Doris Day 1953, coloring/paper doll book
2107	Peek-A-Boo 1955
2108	Ava Gardner 1953
2109	Here's the Bride! 1953
2110	The Christmas Story 1952, stand-ups
2111	Howdy Doody Puppet Show 1952, six puppets to assemble
2112	Walt Disney's Peter Pan Punch-Out Book 1952, stand-ups
2112	Elizabeth Taylor 1954
2113	Spaceport U.S.A. 1953, stand-ups
2114	Goldilocks and The Three Bears 1953, stand-ups
2115	Halloween 1953, party items to assemble
2115	Halloween Party Book 1954, party items to assemble
2116	Lucille Ball, Desi Arnaz with Little Ricky 1953
2117	Giant Model Book 1937, stand-ups
2117	Lullaby Cut-Out Dolls 1953
2118	Roy Rogers, Dale Evans 1953
2119	Boys and Girls Doll Book 1955
2129	Betty and Billy 1955
2131	Mommy and Me 1955
2132	Tiny Tots 1956
2133	Here's Trixie Belden 1956
2135-12	Barbie Sticker Fun 1989, includes paper doll
2135-16	Barbie Sticker Fun 1989, includes paper doll (2135-12)
2139	Campus Queens 1957
2145	3 Little Girls 1957
2148	The Paper Doll Cut out Box 1939, contains the five #1002 books
2160	Playtime Puppet Theater, Mr. Rogers' Neighborhood 1974, puppets & theater
2169	Howdy Doody Fun Book 1951, activity book with some paper dolls
2170	Simple Objects Sticker Fun 1957, includes paper dolls
2172	The Hair Bear Bunch 1972, stand-ups
2173	Roy Rogers and Dale Evans Punchout 1954, stand-ups
2174	Circus Punchout Book 1954, stand-ups
2177	Peter Rabbit Sticker Fun 1960, includes paper doll
2179	Peter Cottontail 1953, coloring book with two paper dolls
2179	Baby Beans 1971, doll on back cover, no outfits
2185	Walt Disney's Snow White and the Seven Dwarfs 1938, box set
2185	Walt Disney's Peter Pan Fun Book 1952, includes paper dolls
2190	The Dionne Quints Cut-Out Dolls 1937, box set (1055)
2192-31	Starr Sticker Fun 1980, has three paper dolls with clothes to color
2192-41	The Littles Dolls 1981, sticker book with seven stand-up dolls, no outfits
2194	Alice in Wonderland 1951, stand-ups
2197	Roy Rogers and Dale Evans 1952, stand-ups
2198	Hopalong Cassidy 1951, stand-ups
2215	Barbie & Gala Evening Fashions 1998
2221	Walt Disney's Mickey and Minnie — Town & Country 1998
2222	Mulan Surprises and Disguises 1998
2302	Walt Disney's Mickey Mouse Decorate-A-

Tree 1979

2302-22 Walt Disney's Mickey Mouse Decorate-A-Tree 1979 (2302)

2303 The Sesame Street Decorate-A-Tree Book 1979

2303-2 The Sesame Street Decorate-A-Tree Book 1979 (2303)

2303-22 The Sesame Street Decorate-A-Tree Book 1979 (2303)

2304 Raggedy Ann And Andy Decorate-A-Tree 1980

2304-1 Raggedy Ann And Andy Decorate-A-Tree 1980 (2304)

2304-22 Raggedy Ann And Andy Decorate-A-Tree 1980 (2304)

2305-1 Walt Disney's Winnie the Pooh Decorate-A-Tree 1980

2305-2 Walt Disney's Winnie the Pooh Decorate-A-Tree 1980 (2305-1)

2305-22 Walt Disney's Winnie the Pooh Decorate-A-Tree (2305-1)

2307-1 Christmas Trims 1980

2308 The Ginghams Decorate-A-Tree 1982

2309 Little Golden Book Friends Decorate-A-Tree 1982

2371 Barbie 1991/94 (1695-1)

2389 Barbie 1992/94 (1690-2)

2391 Storyland Fun - Playskool Activity Book 1989, with paper dolls

2625 Dimple Triplets box, no date, 1950's

2627 Playmates box, no date, circa 1953

2627 Teen Time box, no date, 1950's, Sue & Cindy

2627 Teen Time Dolls box, no date, circa 1954, Peggy & Ann

2630 Pat, A Fabrik Doll Kit box, no date, circa 1953

2630 Terry, A Fabrik Doll Kit box, no date, circa 1953

2748 Barbie 1993/94 (1502-4)

2904 Pinocchio, A Puppet Show with Stage, 8 Characters pictured in 1940 catalog

2920 The Busy Busy Box Contains four activity books. Two contain paper dolls: Let's Make Something 1953, & Let's Have Fun 1954

2925 Bugs Bunny on the Farm 1983, stand-ups

2926 Going for a Ride In the Country, In the City stand-ups

2927 Walt Disney's Winnie-the-Pooh and Friends stand-ups

2928 Walt Disney's Mickey and Minnie Out West 1983, stand-ups

2945 2 Wood Dolls box, date n/a, pictured in 1950 catalog

2950 30 Toy Soldiers stand-ups, pictured in 1943 catalog

2956 Let's Make Something 1953, includes paper doll (fewer pages than #2920)

2956 Let's Have Fun 1954, includes paper doll (fewer pages than #2920)

2964 4 Movie Starlets pictured in 1943 catalog, box (991)

2966 3 Young Americans box 1943

2968 College Girls 1943 box (2966)

2971 Sugarplum Dolls 1957 box

2988 3 Wood Dolls box (982 & 998)

2990 3 Wood Dolls date n/a, pictured in 1948 catalog, box

3005 Walt Disney's Snow White 1938, box

3007 Fifty Little Tots for Little Girls and Boys no date, box, stand-ups

3013 Twelve Stand-up Dolls - Stand-Up Toys no date, 1920's-30's, box

3032 Dolls of Many Lands box (3046)

3036 Whopper Cut Out Doll Box (920, 965, 983 & 997)

3037 Patsy A Wooden Doll box (#1055 Annette for clothes, doll new)

3037 2 Large Dolls, Peggy and Peter box (965)

3044 Polly and Peter and Four Little Friends box contained first three 3044 books listed below

3044 Jackie and Joan box with last three 3044 books listed below

3044 Polly and Peter 1933 (917)

3044 Davy and Dorothy 1933 (917)

3044 Nancy and Peggy 1933 (917)

3044 Tommy and Jean 1933 (917)

3044 Jackie and Joan 1933 (917)

3044 Barbara and Baby Betty 1933 (917)

3046 Dolls of Many Lands 1931, box

3046 Dolls of Many Lands 1932, box

3057 The House That Jack Built in 1933 catalog, stand-ups

3059 4 Dolls 1933 (920)

3063 Dolls and Dresses, The Big Little Set circa 1930's (917, 920 & 991)

3075 Two Large Dolls reprint of 911 with redrawn dolls

3081 10 Dolls from as Many Lands 1934, box

3082 Four Happy Dollies 1933, box (920)

3082-D Funtime Busy Book 1963, with paper dolls

3083 Buddy and Judy Paper Doll Coloring Box 1934 (2021)

3084 Five Little Friends 1933, box, dolls are from #917 books & have clothes to be colored

3113-86 Punky Brewster 1986, coloring/paper doll book

3175 Barbie 1988, coloring/paper doll book

3175-2 Barbie 1988, coloring/paper doll book (3175)

3193 What's In My Basket? 1990, includes paper dolls

3336 Disney's Aladdin 1992, coloring/activity book with paper dolls

3911 125 Play Cut-Outs box of stand-ups from the five #1054 books

3925 Peggy and Peter 1935, box (965)

3926 Baby Betty Doll House with Furniture 1939, box set includes some dolls, outfit availability unknown

3945 American Nurse, A Wood Doll 1941, envelope

3946 American Airlines Stewardess, A Wood Doll 1942, envelope

3951 Punch and Judy Puppet Show in 1940 catalog, box with six puppets & stage

3967 A Wood Doll, Sally 1941, box

3967 A Wood Doll, Joan 1941, box

3967 A Wood Doll, Mary 1941, box

3967 A Wood Doll, Lois 1941, box

3967 A Wood Doll, Betty 1941, box

3967 Patsy, A Wooden Doll box (1055 Annette, clothes, doll new)

3967 Dotty, A Wooden Doll box (1055 Marie, clothes & doll new)

3967 Peggy, A Wooden Doll box (1055 Yvonne, clothes & doll new)

3967 Margie, A Wooden Doll box (1055 Cecile, clothes & doll new)

3967 Bunny, A Wooden Doll box (1055 Emilie, clothes & doll new)

3980 Our WAVE Patsy 1943, box or envelope

3980 Our WAAC Joan 1943, box or envelope

3980 Our Nurse Nancy 1943, box or envelope

3980 Our Soldier Jim 1943, box or envelope

3980 Our Sailor Bob 1943, box or envelope

3983 Big Big Doll 1943, box or envelope

3984 Baby Dolls 1943, box or envelope

3985 A Wood Doll, Babs no date, in 1947 catalog, box or envelope

3985 A Wood Doll, Bunny no date, in 1947 catalog, box or envelope

3985 A Wood Doll, Sissy no date, in 1947 catalog, box or envelope

3985 A Wood Doll, Penny no date, in 1947 catalog, box or envelope

3985 A Wood Doll, Ginger no date, in 1947 catalog, box or envelope

3988 Nursery Dolls 1943, box (3984)

3989 Twin Baby Dolls box (two dolls from #989 6 Cut-Out Dolls)

3995 Peggy and Peter 1935, box (965)

4027 Read, Color, Play 1954, includes paper dolls

4041 Let's Make Something 1953, includes paper dolls

4081 Big and Little Dolls 1954

4098 Contemporary Playhouse 1978, includes paper dolls

4099 The Country Playhouse 1978 includes paper dolls

4100 A Wood Doll, This is Nancy no date, in 1948 catalog, box

4100 A Wood Doll, This is Pat no date, in 1948 catalog, box

4100 A Wood Doll, This is Susan no date, in 1948 catalog, box

4100 A Wood Doll, This is Margie no date, in 1950 catalog, box

4100 A Wood Doll, This is Betty no date, in 1950 catalog, box

4100 A Wood Doll, This is Sandy no date, in 1950 catalog, box

4105 Brother and Sister Dolls no date, in 1949 catalog, box

4112 Peggy and Peter no date, in 1950 catalog, box

4116 A Wood Doll, Penny no date, circa 1950's, box

4116 A Wood Doll, Sunny no date, circa 1950's, box

4123 The Roly-Poly Twins no date, in 1952 catalog, box

4124 Penny, A "Sweetie" Doll no date, circa 1953, box

4124 Polly, A "Sweetie" Doll no date, circa 1953, box

4124 Sweetie Doll, Honey no date, circa 1953, box

4124 Sweetie Doll, Sugar no date, circa 1953, box

4125 The Polka Dot Tots no date, circa 1950's, box

4129 Tiny Tots (Betty & Bill) no date, circa 1954, box

4129 Tiny Tots (Freddy & Joyce) no date, circa 1956, box

4129 Sonny and Sue, Tiny Tot Dolls no date, circa 1954, box

4129 Gerry and Ginger, Tiny Tot Dolls no date, circa 1954, box

4155 A Teeny Tiny Paper Doll 1967

4195 This Is Lisa 1978, box

4196 This Is Sue Ann 1978, box

4197 This Is Marcie 1978, box

4198 This Is Katie 1978, box

4212 The Ginghams: Becky's School Room 1978, box

4214 The Ginghams: Katie's Country Store 1978, box

4215 The Ginghams: Sarah's Farm 1978, box

4216 The Ginghams: Carrie's Birthday Party 1978, box

4217 The Ginghams: Sarah's Picnic 1976, box

4218 The Ginghams: Carrie's Bedroom 1976, box

4219 The Ginghams: Becky's Playroom 1976, box

4220 The Ginghams: Katie's Ice Cream Parlor 1976, box

4296 Play Airport 1975, box, stand-ups

4297 Play Seaport 1975, box, stand-ups

4300 Poochie Play Set 1983, box, includes doll outfits

4301 The Berenstain Bears 1983, box, includes doll outfits

4303 The Berenstain Bears Treehouse Playscene 1983, box, includes outfits

4305 Play Indian Village 1974, box, stand-ups

4306 Play Barnyard 1974, box, stand-ups

4308 A Visit To Walt Disney World 1971, box, stand-ups

4310 Archie Gang At Pop's 1970, activity box, includes paper dolls

4312 Donald Duck Visits Disneyland no date, circa 1970, box, stand-ups

4317 Funtime Puppets 1976, box, puppets

4319 Raggedy Ann and Andy 1973, box

4320 The Brady Bunch 1973, box

4321 Walt Disney's Snow White and The Prince 1967, box (4732)

4322 Barbie Magic Paper Doll 1973, box

4328 Barbie Design-A-Fashion Paper Doll Kit 1979, box

4328-20 Barbie Design-A-Fashion Paper Doll Kit 1979, box (4328)

4328-21 Barbie Design-A-Fashion Paper Doll Kit 1982, box, different from above

4329 **Denim Deb's Design-A-Fashion Paper Doll Kit** 1979, box

4329-21 **Skipper Design-A-Fashion Paper Doll Kit** 1982, box

4330 **Dawn Magic Paper Doll** 1971, box

4331 **Barbie Magic Paper Doll** 1971, box

4332 **New 'N' Groovy P.J. Magic Paper Doll** 1971, box

4333 **Miss America** 1974, box

4334 **The Waltons** 1974, box

4335 **Baby Tender Love** 1974, box

4336 **Newport Barbie and Ken** 1974, box

4337 **The Sunshine Family** 1974, box

4338 **Sun Valley Barbie and Ken** 1974, box (dolls same as 4336, clothes new)

4339 **Mrs. Beasley** 1974, box

4340 **The Brady Bunch** 1974, box (4784)

4340 **Whirlikids Carousel Action** Tammi & Tanzi Toe, 1971, box

4341 **Whirlikids Four Seasons** Merry & Penny Weather, 1971, box

4341 **Mickey and Minnie Mouse** 1972, box (4636)

4342 **Whirlikids Lollipop Action** Twixie & Twirly Pop, 1971, box

4343 **World of Barbie Play Fun Box** 1972, includes paper dolls

4345 **Bugs Bunny and His Friends Puppet Plays** 1972, box, puppets & stage

4347 **Barbie Country Camper and Paper Dolls** 1973, box

4352 **Busy Town** 1972, box, stand-ups

4375 **Wedding Bell Paper Dolls** 1971, box

4376 **World Of Barbie Paper Dolls** 1972, box

4378 **Walt Disney's Storytime Puppets** 1977, box, puppets & stage

4380 **Scissors Snip-Its** 1975, box, stand-ups

4381 **Playtime Zoo** 1975, box, stand-ups

4382 **Land of Dinosaurs** 1975, box, stand-ups

4388 **Shirley Temple** 1976, box

4389 **Barbie and Ken All Sports Tournament** 1976, box

4390 **Pippi Longstocking** 1976, box

4391 **Ballerina Barbie** 1976, box

4392 **Barbie and Francie** 1976, box

4393 **Malibu Francie** 1976, box

4394 **Dusty** 1974/76, box (doll same as 1991, clothes new)

4395 **Skipper** 1976, box

4396 **Raggedy Ann and Andy** 1975, box

4397 **Drowsy** 1975, box

4398 **Baby Alive** 1973/75, box

4399 **Quick Curl Barbie** 1975, box

4401 **Teen Time** Jill & Joan, no date, circa 1956, box (974)

4401 **Teen Time Dolls** Kay & Kim, 1958, box

4401 **Teen Time Dolls** Pat & Carol, 1959, box

4401 **Teen Time Dolls** Merri & Ginny, 1960, box

4401 **Amy Magic Doll** 1965, box

4401 **Sunny Magic Doll** 1966, box

4411 **Big and Little Sister** 1962

4412 **Tiny Tots** Kay & Jay, 1957, box (4123)

4412 **Tiny Tots** Barb & Robby, 1959, box

4412 **Tiny Tot Dolls** Kip & Sis, 1962, box

4415 **Baby Doll** 1962, box

4415 **Playmate** 1965, box

4418 **Tiny Tot Dolls** Kay & Skippy, no date, circa 1960, box (4123)

4418 **Tiny Tot Dolls** Kip & Sis, no date, circa 1960's, box (4412)

4418 **Tiny Tot Dolls** Kay & Jay (4123)

4574 **Family of Dolls** 1960, box (6 dolls from 5 box sets)

4601 **Barbie** 1963, box

4601 **Heidi** 1967, box

4605 **Barbie Wedding Dress 'n Fashion Clothes** 1964, box

4605 **Barbie Fashion Window Wardrobe** 1965, box

4607 **Miss Teen Cut Outs** Cindy, 1960, on cardboard with transparent wrap

4607 **Miss Teen Cut Outs** Kitty, 1960, on cardboard with transparent wrap

4607 **Miss Teen Cut Outs** Vicky, 1960, on cardboard with transparent wrap

4607 **Miss Teen Cut Outs** Wendy, 1960, on cardboard with transparent wrap

4607 **Skipper Day-by-Day Wardrobe** 1965, box, 1964 on doll

4607 **Skipper Fashion Calendar Wardrobe** 1965, box

4607 **Baby PeeWee** 1968, box

4609 **Patty Duke Fashion Wardrobe** 1965, box

4610 **Lucy** (Lucille Ball) 1963, box

4612 **National Velvet** 1962, box

4612 **Bride and Groom** 1966, box

4612 **Bridal Doll, box** 1968, box

4613 **Janet Lennon** 1962, box

4613 **Gretchen** 1966, box

4613 **Trisha** 1968, box

4614 **Connie Stevens** 1961, box

4616 **Bunny Lou** 1960 on cardboard w/transparent wrap

4616 **Mary Lu** 1961 on cardboard w/transparent wrap

4616 **Barbie Travel Wardrobe** 1964, box

4618 **Magic Stay-On Dresses** Julie, 1961, box (4626)

4618 **Magic Stay-On Doll** Kelly, 1963, box

4618 **Sue the Magic Doll** 1964, box

4618 **Vicki Paper Doll** 1966, box

4618 **Amy Magic Doll** 1968, box

4620 **Tammy School and Sports Clothes** 1964, box

4620 **Tammy A Closet Full of Clothes** 1965, box

4620 **Secret of Nimh** 1982, box, stand-ups

4621 **Annette** 1962, box

4621 **Mary Poppins** 1964, box (4702)

4621 **Mary Poppins** 1972, box (4702)

4621 **Mary Poppins** 1966, box, different from above

4622 **Tutti, Barbie & Skipper's Tiny Sister** 1967, box

4624 **Here's Mary with Real Golden Curls** no date, circa 1956, box

4624 **Peggy Has Real Curls** 1957, box

4624 **Karen Has Real Curls** 1958, box

4624 **Stephanie the Teenage Model** 1969, box

4625 **Bridal Party** no date, circa 1955, box

4625 **Bridal Party** no date, circa 1957, box

4625 **Bridal Party** 1959, box

4625 **Baby Sister & Me** 1969, box

4626 **Magic Stay-On Dresses for Patsy** no date, circa 1955, box

4626 **Magic Stay-On Dresses** Sally, no date, circa 1956, box (doll from #2945)

4626 **Magic Stay-On Dresses** Susie, no date, circa 1957, box

4626 **Magic Stay-On Dresses** Mary Anne, no date, circa 1958, box

4626 **Magic Stay-On Dresses** Julie, 1959, box

4626 **Magic Stay-On Dresses** Carol, 1960, box

4626 **Tammy Date and Dress-up Clothes** 1964, box

4626 **Rag Doll** 1969, box

4627 **Mickey Mouse and Friends Fun Box** 1981, puppets, masks, etc.

4636 **Mickey and Minnie Mouse** 1972, box

4639 **Scooter Fashion-Go-Round** 1965, box

4640 **Pepper Around-the-Clock Wardrobe** 1965, box

4643 **Baby's Hungry** 1967, box

4671 **Mod Missy** 1969, box

4672 **Eloise** 1969, box

4673 **Daisy Darling** 1969, box

4683 **Teddy Bear Sew-Ons** 1978, box

4684 **Rag Doll Sew-Ons** 1978, box

4684-20 **Rag Doll Sew-Ons** 1978, box (4684)

4687 **Sleeping Beauty** 1967, box, some boxes not dated

4688 **The Aristocats Play Box** 1970, includes stand-up dolls but no outfits

4701 **Sally and Susie Magic Dolls** 1961 (combined boxes of #4626 Sally & Susie)

4701 **Magic Stay-On Paper Dolls** Judy and Joan, 1963, box

4701 **Magic Stay-On 2 Paper Dolls** Sara and Pat, 1964, box

4701 **Magic Stay-On 2 Paper Dolls** Terry and Holly, 1966, box

4701 **Dolly Dears** 1967, box

4701 **Barbie** 1967, box (4785)

4702 **Mary Poppins 3 Magic Dolls** 1964, box

4704 **Twiggy Magic Paper Doll** 1967, box

4708 **My Dolls Take a Trip** 1957, suitcase style box

4711 **Wee Bedtop Village** 1971, box, stand-ups

4712 **Alice in Wonderland** 1972, box

4716 **Liddle Kiddles Play Fun** 1968, box, stand-up dolls, no outfits

4718 **Lennon Sisters** 1959, canister style box

4718 **Lennon Sisters** 1960, canister style box, different from above set

4718 **Malibu P.J.** 1972, box

4719 **Liddle Kiddles Doll Box** 1967

4719 **Ballet Paper Dolls** 1968, box

4719 **Angelique** 1971, box

4720 **Snow White and the Prince** 1967 (4732)

4721 **Pepper and Dodi Garden of Fashion Wardrobe** 1965, box

4721 **Cinderella and the Prince** 1972, box

4722 **Skediddle Kiddles** 1968, box

4723 **Sleeping Beauty Dolls** 1958

4723 **Wedding Day Dolls** 1961, box, dolls same as #1969 Julie's Wedding, clothes new

4723 **Bridal Doll Box** 1963

4723 **Bridal Doll Box** 1964

4723 **Bridal Party** 1966, box

4723 **Bridal Party** 1968, box

4723 **Happy Bridal Paper Dolls** 1971, box

4732 **Little Ballerina Doll Kit** 1961, canister style box

4732 **Snow White and The Prince** 1967, box

4733 **Walt Disney Starlets** Alice, Annette, Cinderella & Snow White, 1960, canister style box

4733 **Tammy and Her Family** 1964, box. Dolls from #1997 but smaller, clothes new

4733 **Growing Sally** 1969, box

4734 **Sabrina** 1971, box

4735 **Barbie** 1971, box

4736 **Walt Disney's Snow White Play Fun** 1967, box, stand-ups

4740 **Cheerful Tearful** 1966, box

4740 **Raggedy Ann and Andy** 1968

4740 **Raggedy Ann and Andy** 1971, same as above, new box cover

4743 **PeeWee** 1966, box

4743 **Archie Paper Dolls** 1969, box

4744 **Betsy McCall** 1971, box

4750 **The Manger Scene** 1959, box, stand-ups

4754 **Winking Winny** 1969, box

4754 **Crissy** 1972, box

4755 **Yogi Bear** 1964, box, stand-ups

4755 **Bambi** 1966, box, stand-ups

4755 **Tippee Toes** 1969, box

4756 **Atom Ant** 1966, box, stand-ups

4756 **Flatsy** 1969, box

4758 **Peepul Pals** 1967, box, four dolls from #1984, clothes new

4760 **Tippy Tumbles** 1969, box

4762 **My Very Best Friend** 1971, box

4763 **Zoo Play Fun** 1966, box, stand-ups

4763 **Barbie** 1969, yellow box cover

4763 **Barbie** 1969, reprint of above with red box cover

4764 **Magic Stay-On Dresses** 1958, box

4764 **Buffy & Jody** 1970, box

4767 **5 Family Affair Paper Dolls** 1968, box

4768 **The Flintstones Play Fun** 1965, box, stand-ups

4770 **Life Size Doll** 1959, on cardboard with transparent wrap

4770 **Little Lulu** 1972, box

4771 **Zorro** 1965, box, stand-ups

4772 **Brigitte** 1971, box

4773 **Green Acres** 1968, box

4774 **Lucky Locket Kiddles** 1968, box

4774 **Beautiful Crissy** 1971, box

4775 **Patty Duke** 1965, box

4776 **Chatty Baby** 1963, box

4776 **Dollikin** 1971, box

4777 **Charmin' Chatty** 1964, box

4778 **Skipper and Scooter Four Seasons**

No.	Description
	Wardrobe 1965, box
4780	Batman Activity, box 1966, stand-ups
4780	Swingy 1969, box
4781	G.I. Joe Activity, box 1965, stand-ups
4784	The Brady Bunch 1972
4785	Barbie and Midge Travel Wardrobe 1965, box
4785	Barbie 1967, box
4786	Baby Nancy 1971, box
4791	Pebbles and Bamm Bamm 1964, box
4793	Barbie, Midge, Skipper 1965, box
4793	Barbie and Francie 1966, box
4795	This is Pennie a "Real Hair" Doll 1962, box
4795	Penny a "Real Hair" Doll 1963 (4795 with new box cover)
4796	The Flintstones 1962, suitcase style box
4796	Pebbles and Bamm Bamm 1965, box
4797	Barbie and Ken Traveling Dolls 1962, suitcase style box
4797	Barbie and Ken 1963, suitcase style box
4798	Lennon Sisters 1962, suitcase style box
4982	Magic Stay-On Dresses Susie 1959, box (4626)
5041-20	Put and Play Space Puppets 1981, puppets with space suits
5168	All By Myself 1987, Sesame St. Get Ready Book, includes paper dolls
5310	Three Little Girls and How They Grew 1943/44, box (979)
5332	The Gang no date, pictured in 1949 catalog, box
5336	Trim Dotty's Dresses no date, pictured in 1949 catalog, box
5338	Make Mary's Clothes 1949, box
5340	Bridal Party Dolls 1949, box
5347	Outdoor Pals no date, pictured in 1952 catalog, box
5349	Bridal Party no date, circa late 1940's-early 1950's, box
5349	Bridal Party no date, circa 1954, box
5368	Circus Queens 1957, box
5500	Jem 1986, coloring/activity book with paper dolls
5501	Jem 1986, coloring/activity book with paper dolls, different from #5500
5502	Defenders of the Earth 1986, coloring/activity book with paper doll page.
5511-2	Disney's Fairy Tales 1987, coloring/activity book with paper doll
5511-3	Disney's Fairy Tales 1987/89/91, coloring/activity book w/paper doll
5522	Barbie 1988, coloring/activity book w/paper doll
5522-1	Barbie 1990, coloring/activity book w/paper doll
5522-3	Barbie 1990/91, coloring/activity book w/paper doll (5522-1)
5525	Bride & Groom 1988, coloring/activity book w/paper dolls
5525-1	Bride & Groom 1988/92, coloring/activity book w/paper dolls (5525)
5535-5	Beauty and the Beast 1993, coloring/activity book w/paper doll
5550	Pretty Pretty Princess 1991
5551	Bride & Groom 1991
5552	Barbie 1991
5553	Full House 1992, box
5554	Cat Woman 1992, box
5555	Fairy Tale Paper Doll 1992, box
5556	Best Friends 1993, box
5559	Barbie 1993, box (1502-4)
5608	Little Travelers circa 1949, suitcase style box
5608	Eight Playmate Dolls box (979 & 3967)
5608	My Dolls Take A Trip circa 1950, suitcase style box
5608	My Dolls Take A Trip Set No. 2 circa 1952, suitcase style box
5608	My Dolls Take A Trip Set No. 3 circa 1953, suitcase style box
5608	My Dolls Take A Trip Set No. 4 circa 1955, suitcase style box
5613	Alice in Wonderland 1951, box, stand-ups
5614	Hat Box Dolls 1950's, box (2627 Playmates)
5614	Hat Box Dolls circa 1952, box (three girl dolls from 5332)
5615	Walt Disney's Snow White and the Seven Dwarfs 1952, box, stand-ups
5624	Lucille Ball–Desi Arnaz and little Ricky 1953, box
6000	Bride Originals Designer Set 1991, box
6001	Star Originals Design Set 1991, box
6002	Fashion Model Originals Designer Set 1991, box
6852	Family of Paper Dolls (six dolls from 997)
6853	Little Brothers and Sisters 1936 (4 dolls from 983)
6854	Baby Nancy and Her Nursery 1937 (938)
6861	Charlie McCarthy (995)
6879	Walt Disney's Pinocchio 1939 (935)
6882	Walt Disney's Pinocchio 1939, stand-ups (974)
6901	New York World's Fair Goodwill Dolls 1939
7316-A	The Ginghams: Becky's Winter Carnival 1979, box
7316-B	The Ginghams: Katie At The Seashore 1979, box
7316-C	The Ginghams: Sarah's Country Kitchen 1979, box
7316-D	The Ginghams: Carrie's Boutique 1979, box
7317-A	The Ginghams: Sarah's Picnic (4217)
7317-B	The Ginghams: Carrie's Bedroom (4218)
7317-C	The Ginghams: Becky's Playroom (4219)
7317-D	The Ginghams: Katie's Ice Cream Parlor (4220)
7319-A	The Ginghams: Becky's Tea Party 1981, box
7319A-20	The Ginghams: Becky's Tea Party (7319A)
7319A-21	The Ginghams: Becky's Winter Carnival (7316A)
7319-B	The Ginghams: Carrie's County Fair 1981, box
7319B-20	The Ginghams: Carrie's County Fair (7319B)
7319B-21	The Ginghams: Katie at the Seashore (7316B)
7319-C	The Ginghams: Katie's Artist Loft 1981, box
7319C-20	The Ginghams: Katie's Artist Loft (7319C)
7319C-21	The Ginghams: Sarah's Country Kitchen (7316C)
7319-D	The Ginghams: Sarah's Pet Shop 1981, box
7319D-20	The Ginghams: Sarah's Pet Shop (7319D)
7319D-21	The Ginghams: Carrie's Boutique (7316D)
7319E	The Ginghams: Becky's School Room (4212)
7319F	The Ginghams: Katie's Country Store (4214)
7319G	The Ginghams: Sarah's Farm (4215)
7319H	The Ginghams: Carrie's Birthday Party (4216)
7381-A	Raggedy Ann and Andy's Fun Filled Playtime Box 1976, includes paper dolls from #1999
7383-A	Mickey Mouse Club Box Of Things To Do 1977, has paper doll book #1979
7386	The First Christmas 1979, nativity scene stand-up box
7407-A	Poochie 1984, box
7407-B	Crystal Barbie 1984, box
7407-C	Rainbow Brite 1983/84, box
7407-D	Walt Disney's Donald and Daisy 1984, box
7407-E	Barbie 1983 (7411-D)
7408-A	Bride & Groom 1979, box
7408-A	Bride & Groom 1982, box
7408A-1	Strawberry Sue 1981, box
7408-B	Fashion Photo Barbie 1979, box
7408-B	Walt Disney's Cinderella and the Prince 1982, box
7408B-1	Barbie 1981, box
7408B-21	Barbie 1981, box (7408B-1)
7408-C	Denim Deb 1979, box
7408-C	Golden Dream Barbie 1982, box
7408C-1	Super Teen Skipper & Scott 1981, box
7408C-21	Super Teen Skipper & Scott 1981, box (7408C-1)
7408-D	Snow White 1979, box
7408-D	Jean Jeans 1982, box
7408D-1	Starr and Shaun 1981, box
7408-E	Malibu Barbie 1982, box
7408-F	Walt Disney's Mickey & Minnie 1982, box
7408-G	Raggedy Ann and Andy 1982, box
7408-H	Western Barbie 1982, box
7409-A	Walt Disney's Donald & Daisy 1979, box
7409-B	Betsy Paper Rag Doll 1979, box
7409-C	Teddy Bears 1979, box
7409-D	Trixie Belden 1979, box
7410-A	Ballet Stars 1980, box
7410-B	Pretty Changes Barbie 1980, box
7410B-20	Pretty Changes Barbie 1980, box (7410-B)
7410-C	Star Princess and Pluta 1980, box
7410-D	Miss America 1980, box
7411-A	The Original Monchhichi Paper Dolls 1983, box
7411-B	Pink & Pretty Barbie 1983, box
7411-C	Western Skipper 1983, box
7411-D	Barbie 1983, box
7413-A	Jean Jeans 1978, box
7413-B	Raggedy Ann and Andy 1978, box
7413-C	Rosebud 1978, box
7413-D	SuperStar Barbie 1978, box
7554A-21	Amy "I Like To Play Dress-Up!" 1981, box
7554B-21	Jennifer "I Like Western Wear!" 1981, box
7554C-21	Jessica "I Like Sports!" 1981, box
7554D-21	Kristen "I Like To Dance!" 1981, box
A7554E-22	Beth "I Like The Farm!" 1982, box
A7554F-22	Julie "I Like The Seashore" 1982, box
A7554G-22	Peggy "I Like The Circus" 1982, box
A7554H-22	Rachel "I Like Make-Believe" 1982, box
7820	Play Pack, Calico Miss Paper Doll 1975, small plastic wrapped package
7821	Play Pack, Calico Miss Paper Doll 1975 (7820)
8070	Pocahontas 1995
8071	Madeline 1996
8073-30	Disney's Hunchback of Notre Dame 1996
8601-30	Disney's Hercules 1997
20275	Pepper Ann 1999
20282	Pooh Let's Dress Up Fun 1999
20283	Barbie Dress Up Fun 1999
21110-00	Barbie, You're the Designer 1999
	The following eight are Gingham's by Rainbow Works
75978-1	Sarah's Country Kitchen (7316C)
75978-2	Katie at the Seashore (7316B)
75978-3	Becky's Winter Carnival (7316A)
75978-4	Carrie's Boutique (7316D)
75979-1	Katie's Country Store (4214)
75979-2	Sarah's Farm (4215)
75979-3	Carrie's Birthday Party (4216)
75979-4	Becky's Playroom (4219)
82539	Barbie & Gala Evening Fashions 1998/2003 (2215)
No #	Barbie Glitter Fashion Maker 1992, cardboard packet/envelope
No #	Cut Out Dolls 1920's, 3 dolls (506 Bobby, Betty and Billie)
No #	Cut Out Dolls envelope with six dolls & clothes
No #	The Dolls That You Love 1910
No #	Jack, Jane and Jerry Cut out Dolls (506)
No #	Little Monster's You-Can-Make-It Book 1978, includes paper dolls
No #	Play Pack & Mini Zoo 1967/75

PHOTO INDEX